Rise to the Occasion

Rise to the Occasion

LESSONS FROM THE Bingham Canyon Manefay Slide

Brad Ross

PUBLISHED BY THE
SOCIETY FOR MINING, METALLURGY & EXPLORATION

Society for Mining, Metallurgy & Exploration (SME)
12999 E. Adam Aircraft Circle
Englewood, Colorado, USA 80112
(303) 948-4200 / (800) 763-3132
www.smenet.org

The Society for Mining, Metallurgy & Exploration, Inc. (SME) is a professional society (nonprofit 501 (c) (3) corporation) whose more than 15,000 members represent professionals serving the minerals industry in more than 100 countries. SME members include engineers, geologists, metallurgists, educators, students, and researchers. SME advances the worldwide mining and underground construction community through information exchange and professional development.

ISBN 978-0-87335-431-8

Ebook 978-0-87335-434-9

Library of Congress Cataloging-in-Publication Data

Names: Ross, Brad, author.
Title: Rise to the occasion : lessons from the Bingham Canyon Manefay slide / Brad Ross.
Description: First edition. | Englewood, Colorado : Society for Mining, Metallurgy & Exploration, [2016] | Includes bibliographical references and index.
Identifiers: LCCN 2016045825 (print) | LCCN 2016051311 (ebook) | ISBN 9780873354318 (print) | ISBN 9780873354349 (ebook)
Subjects: LCSH: Copper mines and mining—Accidents—Utah—Salt Lake City Region. | Landslides—Utah—Salt Lake City Region. | Copper mines and mining—Safety measures. | Bingham Copper Mine (Utah)—Accidents.
Classification: LCC TN317 .R67 2016 (print) | LCC TN317 (ebook) | DDC 363.11/96223430979225—dc23
LC record available at https://lccn.loc.gov/2016045825

Dedication

To the men and women of the Bingham Canyon Mine who rose to the occasion.

Your resourcefulness has inspired me,

Your dedication has energized me,

Your leadership has empowered me,

And your support has humbled me.

This book is for you. Your story is one that should be told

and your accomplishments remembered.

Contents

6880 Complex
Fortuna Underweighting
Head Scarp
Intermediate Scarp
10% Debris
E3 Bench Clearing
Keystone Access
Cornerstone
Manefay Failure Area
6190 Knob
E5
10% Hail Road
Crusher and Conveyor
Tunnel
E4 South
Uncover Next Ore
6190 Complex
10% Slot and Fill
Moly Dome
Underground Facilities
(Northwest Drainage Gallery)
Fill Bridge 2
Main Hill

Preface

This is the true story about the Manefay highwall failure that occurred at the Bingham Canyon Mine outside of Salt Lake City, Utah, on April 10, 2013. The Bingham Canyon Mine is part of Kennecott Utah Copper LLC, which in turn is owned by Rio Tinto. The Manefay was the largest highwall failure (landslide) in mining history at the time of writing this book. Being the largest in history, the Manefay was big—really big. A total of 66.2 million cubic yards, or 144.4 million short tons, fell more than 2,000 feet and traveled over 1.5 miles in two events. The episodes occurred nearly 90 minutes apart, and each one lasted approximately 90 seconds. To put this in perspective, this volume of material is approximately equal to 20 Dallas Cowboy Stadiums (the largest stadium by volume) filled with solid rock dropping from a height of the second tallest building in the world, the 128-story Shanghai Tower.

As large as the Manefay failure was, the really amazing statistic is *zero*—there were zero injuries and zero fatalities. Unlike mining disasters that fill the news media for weeks at a time and where miners are trapped or killed, the Manefay had little coverage—just a few news stories with impressive pictures and a short report of what happened. This modest reporting was probably because the Manefay was a crisis and not a disaster. And it was only a crisis because the right decisions were made to keep people safe and return the mine to production quickly and efficiently. Although this does not necessarily make for the best news item, it is a compelling story of leadership, innovation, and teamwork.

My connection with the Bingham Canyon Mine and the Manefay started on December 1, 2012, when Matt Lengerich, the relatively new general manager of the Bingham Canyon Mine, was sitting at the kitchen counter at my home in Tucson, Arizona. Matt was in town to receive the prestigious Medal of Merit Under 40 from the Mining Foundation of the Southwest and had stopped by the house to say hello before the event. Matt is a good friend and we had worked together at the Colowyo Coal Mine in Colorado in 2002 where he was building a process improvement team.

During our discussion, Matt started to question me on how I would set up the Technical Services Department at the Bingham Canyon Mine and then described how he had some really good technical employees, but they needed a bit of guidance and development to prepare them for future positions. Matt knew that I could not resist the opportunity to help develop talented people, so we started a lengthy negotiation process for me to move to Salt Lake City and manage the mine's Technical Services Department.

As with many transfers, there seemed to be no end to the little details that had to be agreed upon to make the move happen. Some days it seemed that we would never come to a resolution. Finally, on February 12, 2013, Matt called me and basically said that all of my demands had been met! It was a bit of surprise, but I was happy to have the negotiations completed, and we decided that I would report to work on March 1, 2013. I would find out much later that February 12 was an important day for another reason—that was the day that the geotechnical team reported that the mine could experience a large highwall failure sometime in the future.

On February 21, Matt called me at 9:30 in the evening—a bit unusual for him to call so late. My wife Linda and I were working in our home office when the call came in, and Matt started out by saying that he needed to tell me that the Bingham Canyon Mine was going to experience a "little" highwall failure. That didn't faze me much, given that I had worked in other mines that had experienced highwall failures—it was just part of the job. When I asked how "little" of a failure, Matt responded with, "Well…it could be tens to a hundred million tons."

"What!?" was my first thought. I had never even heard of a highwall failure anywhere near a hundred million tons. A *million tons* is a very large failure—*100 million tons* is almost unbelievable. Matt followed up by saying that the geotechnical engineers "were not sure how large it was going to be, but it would not happen right away. It could be several weeks or probably several months away."

After taking a few seconds to process the enormity of the information, my response was decisive, "Matt, I can be there tomorrow."

After a short hesitation, Matt replied, "No, no. That's alright…. We could maybe use you, but … we can get by…"

Matt and I talked a bit more and agreed that I would be at the mine on the scheduled date of March 1. Then we said goodbye and hung up.

Afterward I again sat in silence for a few seconds, thinking about the short discussion. My wife had overheard the conversation and, as usual, read my mind asserting, "You have to go tomorrow." To which I acknowledged, "Yeah, I've got to go tomorrow."

It was not yet 10:00, so after making a few quick arrangements, I picked up the phone again and called Matt. When he answered, my first words were, "I can be there tomorrow *morning*," to which Matt exclaimed, "Great! I will see you in the morning!" Matt would later tell me that it was the first good night's sleep he'd had in several nights.

By 11:30 a.m. the next morning, I was walking through the doors of the Administration Building at the Bingham Canyon Mine, and so began the two most intense but amazing years of my career. During that time, I was able to be a part of a company that prepared for, experienced, and recovered from the largest mining landslide in history. I witnessed tremendous dedication, motivation, innovation, and leadership from people at all levels of the company, as well as contractors and vendors, as everyone worked to keep people safe and return the mine back to normal operations. The hours were long and the pressure to save the mine was great, but the wisdom gained and satisfaction of being part of the team that would "Rise to the Occasion" and save the historic mine were even greater.

In bygone days, great deeds and heroic efforts were retold in stories that were passed down from generation to generation. These stories were shared around a campfire or at the kitchen table so others could learn from the events that transpired. To me, the Manefay slide at the Bingham Canyon Mine was an epic adventure where the heroes were the mine employees, contractors, and vendors working together to save the company, and the villain was the Manefay, the monster landslide that threatened to bring about death and destruction. It was an event worthy of a story, not only because of the battle between heroes and villain, but also because of the tremendous lessons learned that can potentially help others to prevent or react to similar events, and even be used to improve everyday work flow and operations. If we could only apply such important knowledge without the monstrous event, imagine what could be achieved.

There are some differences between *Rise to the Occasion—Lessons from the Bingham Canyon Manefay Slide* and the stories of old, the first being that all these events are true and as factual as possible. The second difference is that unfortunately, I cannot tell the story of the Manefay as we sit around a campfire—but at least I can write down the facts of the Manefay so you can read, enjoy, and perhaps learn at your leisure.

This chronicle begins in the middle of the event, on the day the Manefay landslide actually happened, and describes the sheer magnitude of the Manefay and the series of events that occurred directly before and after the slide. In Chapter 2, the story goes back to before the landslide and describes how the Manefay was detected and what was done to keep people safe. Chapters 3 and 4 then relate how the company safely and quickly recovered from the Manefay events and returned to normal production. Chapter 5 discusses how the company applied lessons from the Manefay to prevent yet another large landslide, and Chapter 6 reveals why the Manefay failed. The final chapter covers innovations and cultural changes that helped the company to be successful in a time of great adversity. The events in this book were large and impressive, but it was the actions of the people involved that made the difference in turning what could have been a truly disastrous event into a story of perseverance and success. Indeed, the lessons learned originated from the people at Kennecott.

I hope you enjoy reading *Rise to the Occasion* and find a few nuggets of wisdom that will help you going forward.

Lessons Learned

Within the story of the Manefay failure are many lessons that can be applied to other businesses or situations. A few of these lessons are very specific to mining or geotechnical issues, such as the use of Voellmy friction factors in dynamic models to predict how extremely large masses of earth like the Manefay might fail. However, a large majority of the lessons can be applied to nearly every business or organization that faces any type of significant risk. Some of these lessons learned are basic and already part of the culture in many organizations, such as knowing what the greatest risks are to the company. But knowing a risk and preparing for it are different concepts. An important lesson from the Manefay is the use of multiple methods to monitor the critical risk and then to act on that monitoring. This method was used by the Bingham Canyon Mine to save lives and prevent injuries.

It could be concluded that the lessons learned are applicable only to potential disasters or crisis situations—how to deal with them and how to avoid them. But the Bingham Canyon Mine also changed its processes and culture as the 100-year-old organization became better, faster, and more innovative—almost overnight. Perhaps there is an opportunity to apply the lessons learned from the Manefay slide to make changes to the culture of your organization—to be better, faster, and more innovative by using what was done in a crisis without having to go through a traumatic event. Therefore, a summary of what I consider to be the key lessons learned from the Manefay failure are included at the end of each chapter so you can apply the lessons to your own situation.

Acknowledgments

Assistance was needed from many people for me to be able to write this book. First, deep appreciation and gratitude go to the management and leadership of Kennecott Utah Copper and Rio Tinto for allowing me to use facts and photographs from the Manefay slide—without your tremendous support, it would not have been possible. Your willingness to share this information speaks volumes about the company's commitment to safety.

Special thanks go to Hugh Thatcher and Cody Sutherlin and so many others who helped review and fact-check the content—it is much improved because of your time and effort.

I am also indebted to Kelly Sanders, Matt Lengerich, and Stephane Leblanc—your leadership provided the environment for all of us to *rise to the occasion*.

To all the men and women of Kennecott, I extend my sincere thanks for taking photographs, generating drawings and statistics, and telling your stories about the Manefay—you are an integral part of this book. I cannot think of any group of people I would have rather been with during that time. It was an honor and a privilege to be a part of this terrific team of people who will always be my heroes.

I also wish to express my appreciation and gratitude to Jane Olivier and the Society for Mining, Metallurgy & Exploration. You saw the potential of this story and readily agreed to publish it. That gave me the confidence to go forward, the freedom to focus on writing, and the drive to finish the work. Having Jane as my publisher and Diane Serafin as my editor made all the difference in the world.

Special recognition goes to my kids, Robert and Karen (Ross) Bakken. Bob, I will always see a younger me in you. You remind me of the importance of being innovative and creating something from nothing; I am better for that. Karen, it is not often that a dad gets to work with his daughter during a crisis such as the Manefay, and it was one of the thrills of my life. You did great work—job well done. I could not be prouder of the two of you.

Finally, I owe my heartfelt thanks and gratitude to my spouse extraordinaire, Linda. You have encouraged me in all that I have wanted to do, no matter where we went, no matter how long the days could be, and no matter what I wanted to accomplish in both my personal and professional lives. You always rose to the occasion. Your support, ideas, and feedback on this book helped make this journey a reality. Thank you. I really do not know what I would do without you.

The Day of the Manefay

Chapter 1

We were ready—ready for what the geotechnical engineers had predicted, less than two short months earlier, to be, by far, the largest landslide in the 107-year history of the Bingham Canyon Mine (Figure 1.1). The Bingham Canyon Mine as well as the concentrator, smelter, and refinery are all part of Kennecott Utah Copper LLC, which is owned by Rio Tinto. The slide could be an order of magnitude larger than anyone at the mine had ever experienced. In the two months prior to this day, the employees and contractors had done an amazing amount of work to prepare for what would be known as the Manefay (pronounced *main-fay*) landslide. Monitoring of the highwall had been increased to detect any unexpected movements or trends so an early warning could be given to keep people safe. Preparations had also been made for the eventual failure by moving buildings and infrastructure, developing a plan for how to react to different stages of the failure, building a secondary access road, and preparing to return to production as soon as possible when the failure was complete.

Figure 1.1 • Aerial View of the Bingham Canyon Mine in Utah, 2012

In prior months, Bingham Canyon Mine had prepared a Manefay Response Plan that provided instructions for what should be done at various stages as the countdown to failure approached. The stages were broken down into color levels and numbered from 0 to 4. An abbreviated version of the plan is shown in Figure 1.2. The highwall movement was closely monitored and communicated to all employees daily. Everyone was keenly aware of the progress and what to do at each level. The first level was green, which meant that mining operations should progress as normal, even though preparations for the failure were feverishly going on in the background.

Response Level	Manefay Timing	Response
0 (Blue)	Stable conditions	Routine operations
1 (Green)	Weeks to several months	Routine operations and prepare for failure
2 (Yellow)	Days to weeks	Modify operations, restrict access, and continue preparations
3 (Orange)	Hours to days	Evacuate failure areas and close access
4 (Red)	Unexpected acceleration or failure	Emergency evacuation and response

Figure 1.2 • Abbreviated Manefay Response Plan

On April 5, the Manefay mass acceleration increased and the mine's response level moved to yellow, meaning that the Manefay failure was estimated to be only days to weeks away. Operations were modified to restrict access to critical areas of the mine, and two shovels were moved out of the pit to help with reestablishing the 10% haul road. Figure 1.3 is an aerial view taken in August of 2012 and shows the locations of the various facilities discussed in this book, such as the 10% haul road and other places in the area of the slide. The 10% haul road was the only access for large equipment into the mine and critical for the long-term production of the mine. A large section of the 10% haul road was expected to be damaged or destroyed by the slide. Figure 1.3 also shows the Moly Dome, which was a 300-foot-high block of very low copper grade, but high-grade molybdenum (moly) rock that had been left in the bottom of the gigantic pit for economic reasons. The belief was that the Moly Dome would block any debris material if the failure was larger than expected. Parts and supplies were being moved to the Moly Dome so that the ore operations could quickly and efficiently resume at the end of the Manefay failure.

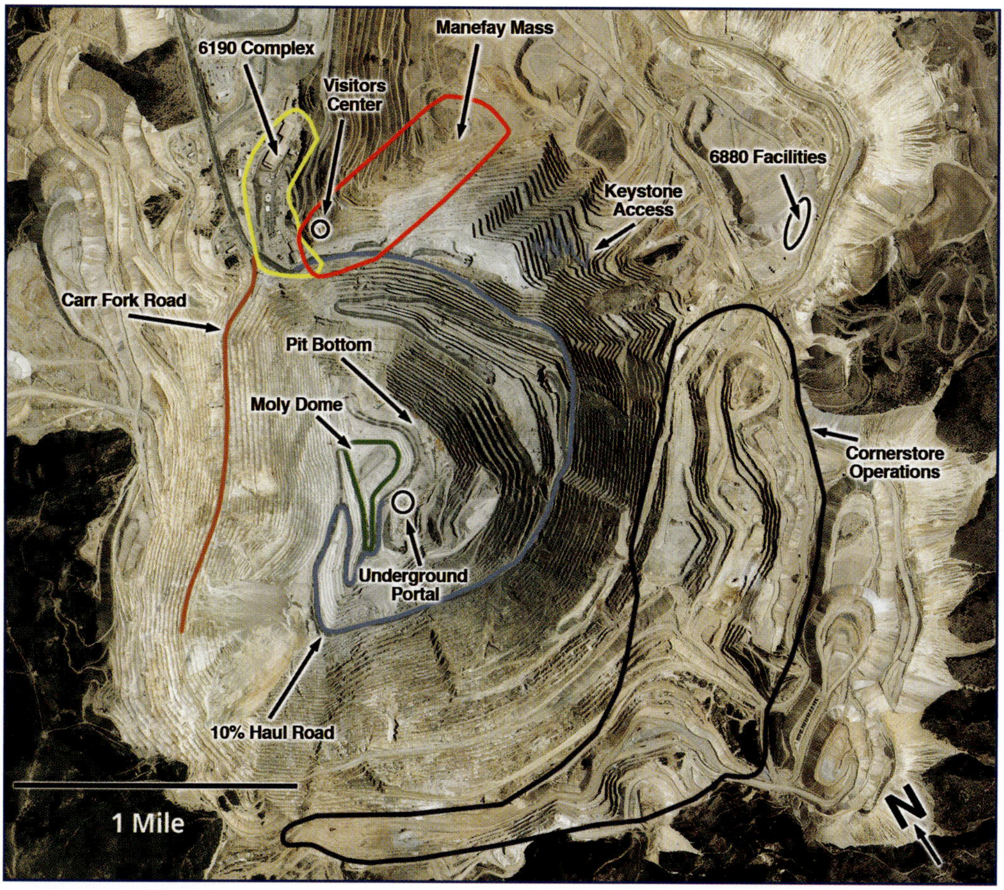

Figure 1.3 • Bingham Canyon Layout in August 2012

Everyone at the mine was anxious and wondering when we would move to Level Orange. At this level, the failure was expected to be only hours to a few days away, and mining operations would be stopped. The 10% haul road would shut down and all equipment would be moved to safe areas well outside the predicted failure zone. The efforts to predict the Manefay and the preparation made before the failure are discussed in detail in Chapter 2.

The attitude of the workforce was of heightened concern—but also confidence. The frequent communications meant that everyone was aware of the movement and everyone could see all of the work that was going on to prepare for the Manefay. People were confident of the skills of the geotechnical engineers to forecast the failure—after all, they had already given the mine significant warning. The speed and amount of preparation was impressive. In just about everyone's mind, we were ready—at least we were ready for what was *expected.*

Making the Call to Level Orange

On April 9, the movement of the Manefay had doubled from 1.3 inches per day to 2.6 inches per day, and significant cracking had occurred in the haul road near the Bingham Shop, indicating a change in the movement of the Manefay mass. The Bingham Shop was a repair and maintenance facility that normally housed tools and mining equipment in various stages of repair but was currently unoccupied and locked shut. The shop was located closest to the gigantic open pit near a group of buildings known as the 6190 Complex, shown in Figure 1.3. By late evening, the Geotechnical Department, outside geotechnical consultant, and Rio Tinto geotechnical expert Zavis (Zip) Zavodni all agreed that the failure could happen by the 14th and believed that by the next morning, April 10, the mine should go to Level Orange—the bottom of the pit would be evacuated and the 10% haul road would be closed.

By 5:30 a.m. on the 10th, Elaina Ware, the operations manager for the mine, had enough information to make the decision to move to Level Orange and begin the process of evacuating the bottom of the pit. According to the Manefay Response Plan, Level Orange indicated that the failure could occur within a matter of hours to a few days, so time was critical. A meeting of the Mine Management Team had been set for 6:30 a.m. to discuss escalating to Level Orange, but Elaina had full authority to make the decision earlier. Consequently, the Operations Team had already begun the evacuation process. Figure 1.4 shows the communication that went out to employees to inform them that the mine had moved to Level Orange, indicating that the Manefay failure could happen within hours to a few days.

Daily Manefay Status Update

Wednesday, April 10, 2013

Current Manefay Response Level is ORANGE

- Current 'Upper Manefay' overall movement rates are **2.6"/day**
- Current 'Lower Manefay' overall movement rates are **2.6"/day**
- Current Movement Trend: Increasing acceleration with localized rockfall
- Actions Required:
  - Close the upper 10% to all traffic
  - Evacuate the lower pit and suspend mining until further notice
  - Close the mine access road from the notch to 6190
    - Route essential traffic through the 4H road
  - Maintain spotters and sentries at key locations
  - Prioritize activities in alignment with Orange response

Figure 1.4 • Daily Manefay Status, Level Orange

Having a response plan in place resulted in a relatively smooth transition to Level Orange. The Operations Team had designated locations to park the trucks and support equipment well outside the estimated impact area of the slide. Days before going to Level Orange, the estimated extent of the Manefay had been staked in the pit by the surveyors as well as a 1,000-foot offset. A berm was constructed at the 1,000-foot offset, and all shovels and drills were moved to locations outside the berm, away from the slide.

In one twist of fate, the last superintendent in the pit, David Lanham, was making a final pass to make sure everyone was evacuated. As he drove by the S97 shovel, a P&H 4100C, he saw that it was just outside the berm of the projected impact area. David wondered whether the shovel should be moved even farther. He asked the superintendent acting as the incident commander in charge of the bottom-of-the-pit evacuation if he should move it a bit farther. The incident commander told David that the shovel was out of the projected impact zone and he should evacuate the area instead of moving the shovel. The critical issue at that point was to evacuate personnel from the mine, which included David. As it turned out, the S97 shovel was caught on the very edge of the Manefay slide. The shovel was so close to the edge that the boom of the shovel was outside of the failure mass, but the rest would be covered by tons of debris. If the shovel had been moved a few hundred feet farther, it may not have been destroyed. However, following a plan to safely and effectively evacuate the pit was critical for keeping people safe—the highest priority. There could have been a thousand small adjustments made, but the plan was based on the best available information for the impact of the Manefay, and the shovel was considered to be clear. Based on that best available data, the right call was made and the shovel was not moved.

Access to the 10% haul road was barricaded to limit entry to the bottom of the pit. Only the Keystone access road (a narrow, light-vehicle road with 13 sharp switchbacks) would be clear of the predicted failure mass. Equipment and supplies were left on the Moly Dome nearly 300 feet above the bottom of the pit so that after the Manefay failed, operations would be able to quickly and efficiently return to production as soon as possible.

At the Administration Building, the Mine Management Team was meeting to go through the rest of the process to see if anything had been missed. Figure 1.5 shows the various buildings located within the 6190 Complex. All nonessential personnel were preparing to be evacuated from this area. One issue that arose was the possibility of a large dust cloud. The mine had experienced dust in some of the much smaller highwall failures in the past, so it had become an issue of concern. If a dust cloud formed after the failure, it could linger in the pit and the 6190 Complex for several days, making the area hazardous for both people and equipment such as computers. A decision was made to have the technical and administrative staff take their computers and surveying equipment with them as they evacuated the 6190 Complex. This turned out to be a great decision, not because of the dust generated by the Manefay, but for the reason that the technical and administrative people would not be able to return to work at the 6190 Complex for several months. However, since they took their computers when they evacuated, they were able to start the mine planning the day after the Manefay failed.

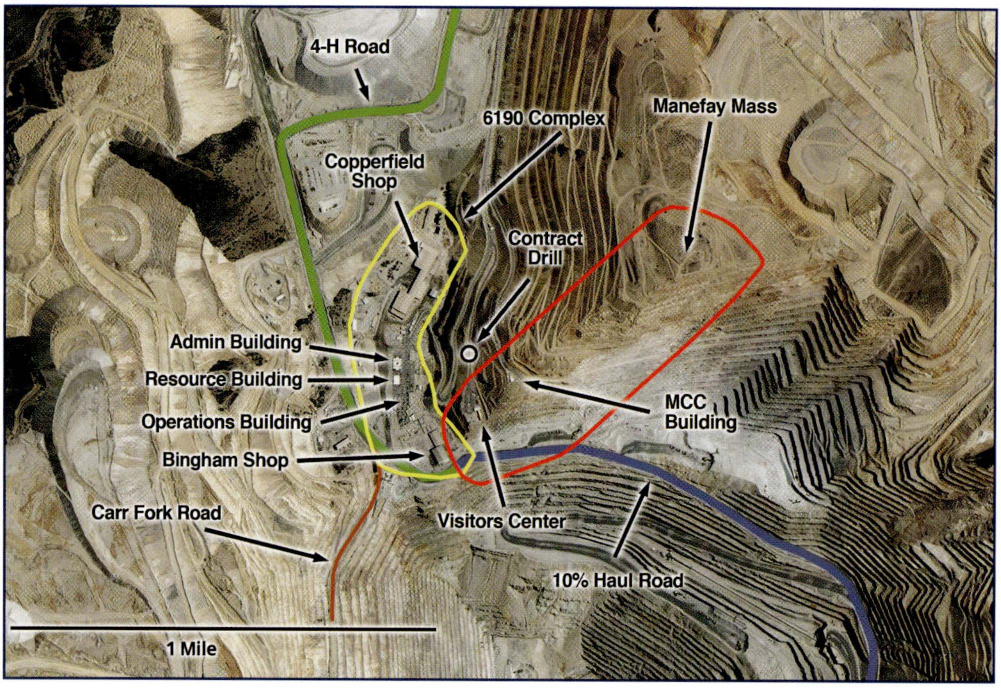

Figure 1.5 • Layout of 6190 Complex

By mid-morning of April 10, most of the administrative and technical staff had been evacuated from the 6190 Complex. Some had moved to the 6880 offices, which were beyond the potential slide zone. Others moved to Rio Tinto's Daybreak offices in South Jordan, approximately 10 miles from the mine. Some employees went home because they did not have access to the work they would normally do.

I spent the rest of the day meeting with the Mine Management Team to ensure that no major issues had been missed in the Manefay Response Plan. I also spent time with Chris Haecker, one of the young mechanical engineers on the Production Support Team, to obtain final approval for the purchase of ten 4×4, high-clearance vans that would be used to transport people in and out of the pit on the Keystone access road after the Manefay slide. We were able to get the final approval from Kelly Sanders, president and CEO of Kennecott, and by the end of the day Chris had all of the vans delivered with all of the site-specific safety equipment installed (radio, lighted flag, fire extinguisher, etc.) so they would be ready when the Manefay did fail. These vans proved to be a critical asset in returning to production after the Manefay because the normal school buses would not have been able to safely move employees in and out of the pit via the sharp switchbacks of the Keystone access road.

During the day, operators continued overburden stripping for the next pushback in the Cornerstone area of the pit. Cornerstone was on the opposite side of the pit as the Manefay and nearly the same elevation as the Manefay failure mass. The access to Cornerstone was not at risk from the impending slide, so normal operations continued.

Critical Communications

The change to Level Orange set a number of communication plans in motion as well. The Kennecott Communications Team sent an internal memo to all employees that the Manefay mass had continued to accelerate and the mine had taken a number of actions to keep people safe. These actions included closing the 10% haul road, restricting access to the mine for all nonessential personnel, relocating people out of affected zones, and activating emergency evacuation plans and teams.

The Bingham Canyon Mine uses many vendors, and a similar communication went out to each of the companies working at the mine through the mine's Operations and Maintenance Teams. These communications were greatly appreciated, as indicated by an e-mail from Nate Kendall, the district manager for Joy Global, to Matt Lengerich, the general manager of the Bingham Canyon Mine:

> I just wanted to drop a quick note to express gratitude for the excellent communications from all of your teams regarding the Manefay failure. The transparency of the information, and all of the incredible work to forecast the slide and the corresponding plan to keep people out of harm's way, has made it possible for all of our team to continue working without being in a confused, rushed, and uncomfortable state.

This e-mail was sent just a few hours after evacuating the 6190 Complex where most of the Joy Global employees worked.

Mr. Kendall's e-mail captured the mood and activities of the morning on the 10th. Everyone went through the process of evacuating and preparing for the Manefay calmly, professionally, and confidently—calmly because they knew they had time to evacuate safely, professionally because there was a plan in place and they were following the plan, and confidently because they trusted the company and the geotechnical information that they were getting. This was in large measure because of the daily information they had been receiving on the Manefay movement for the previous few weeks.

Additional communications went to the Mine Safety and Health Administration (MSHA) that morning as a part of moving to Level Orange. MSHA is the federal agency responsible for enforcing federal health and safety rules and regulations that apply to mines. The principal advisor for safety and health for the Bingham Canyon Mine was the primary MSHA contact for the company. He had been providing updates to the local MSHA inspectors about the Manefay slide and called to let them know that the mine had evacuated the pit bottom because of the imminent slide. This was a rather unusual situation because in most circumstances MSHA would not be notified until after an event happened—often when a fatality or serious injury had occurred. In this situation, however, the company was being proactive and informing MSHA that an event would occur and the company had removed people from harm's way. Although this would not change the fact that MSHA would soon exercise its jurisdiction to protect the safety and health of miners, and issue a Section 103(k) Order (an accident and rescue/recovery procedure) to the Bingham Canyon Mine, it did give the company credibility that it was indeed concerned about the health and safety of its employees.

The Communications Department also started a process of informing the press and government officials that the slide would happen in the relatively near future. These communications before the event prevented key stakeholders from being surprised by the ultimate Manefay slide and also built up the company's credibility.

The open and transparent communications that went on before the Manefay were key to the company being prepared for the ultimate slide. In many ways it would have been easy not to share these communications for fear that the company would look weak. In reality it was important that as many people knew about the slide as possible so they were prepared and the company could meet its primary objective—to keep people safe. The fact that these communications built the trust and respect of employees, contractors, regulators, the press, government agencies, and key stakeholders was a tremendous added benefit that would help the company when it started to recover from the Manefay slide.

Buildup to the Manefay

By the afternoon of April 10, the bottom of the pit had been evacuated and everyone was accounted for. The only people remaining at or near the 6190 Complex were the essential employees, which included members of the Incident Command (IC) Team who were responsible for safely evacuating the mine, full-time spotters who were located at various areas of the mine, dispatchers in the Resource Building, and a geotechnical engineer in the Administration Building who was monitoring the movement of the Manefay mass. Operations were ongoing in the Cornerstone area but would be shut down later in the evening.

A few days before the slide occurred, a contract drill was brought on-site to drill angled holes from behind the Manefay mass to penetrate the mass and the Manefay bed. The plan was to install instruments to detect movement and water to better characterize the potential failure. This information would not only help determine when the Manefay would fail, it would provide critical information on the size and even how the Manefay would fail.

The drill had only been operating for a few days and was close to reaching the Manefay bed when the mine moved to Level Orange. The drill was set up on the road going to the Visitors Center and was behind the Manefay mass. The location was close to the Manefay mass but behind the projected impact area and away from the direction of movement. The decision was made to let the drillers continue to operate in hopes of gathering the critical data on the bed of the Manefay.

Around 8:00 p.m., Collin Roberts, a supervisor, was traveling around the pit to check on the Cornerstone operations. As Collin was passing through one of the checkpoints that had been set up to control traffic going to the mine, he asked the employees working there if they had checked on the contract drillers. The employees said they had and stated that they seemed a little nervous working in that area. Collin gave instructions to have the drill contractors prepare to leave the area. Collin then traveled to the Cornerstone area and picked up Joe Shafer, another supervisor, for a quick inspection of the vicinity.

A short time later, James Jung, a geotechnical engineer, called Collin on the radio—another alarm had sounded and the movement had increased to a rate of 15 inches per day. James was certain that the failure was imminent based on the rapid acceleration.

Collin and Joe headed back to the Carr Fork Road lookout to inspect the highwall. On the way, they called the employees at the roadblock and instructed them to evacuate the drillers—now. Very shortly afterwards, the roadblock personnel radioed back—the drillers were evacuated and accounted for.

MAYDAY! MAYDAY! MAYDAY!

It was dark when Collin and Joe were returning to the Carr Fork Road lookout and they stopped at the 6190 Complex to pick up a strong spotlight to inspect the highwalls. They arrived at the lookout a little after 9:00 p.m. and there were already two spotters there, Nick Patrick and Scott Edwards.

As the four men talked, they could hear rocks and detrital material falling down the highwall. Then at 9:30 p.m. on April 10, 2013, it happened—the Manefay mass literally exploded out of the highwall. It did not fail slowly over hours or days, as previous Bingham Canyon Mine failures had. In 90 short seconds (Pankow et al. 2014), a large section of the Bingham Canyon Mine highwall went from a solid wall of stone to an exploding mass of rock traveling a distance of 1½ miles and dropping thousands of feet as it traveled to the bottom of the Bingham Canyon Mine.

Although it was dark, there was enough light from the facilities at the 6190 Complex that Collin, Joe, Nick, and Scott could see the entire Manefay mass flow down into the black of the pit. As the mass was rushing down, there were large arcs of light from the destruction of substations in the path of the failed Manefay mass. An incredible event to witness!

As the Manefay mass failed, the lights across the mine went out. According to the Manefay Response Plan, the mine was prepared for the Manefay failure, so this was not an unexpected event, and therefore an Incident Command would not have been required. But since the Manefay did not act as expected, Collin and Joe declared an emergency. Collin got on the company radio and shouted, "MAYDAY! MAYDAY! MAYDAY! The Manefay has just failed!"

The results and damage caused by the Manefay would not be completely known until sunrise. Figure 1.6 shows the extent of the failure area as well as the distance that the Manefay traveled before finally stopping after covering the bottom of the pit. As with all declared emergencies at the mine, all operations ceased throughout the pit and would not resume until permission would be given by the dispatchers or supervisors.

Figure 1.6 • Aftermath of the Manefay Failure

The radios and network systems used by the geotechnical monitors had battery backup and were operational. Cell phone service was also functioning because of the decision to relocate the cell phone towers before the Manefay failed—otherwise the mine could have been without communications at that point. Collin, Joe, and the dispatchers quickly started to call superintendents, managers, and the general manager to inform them that the Manefay had failed. As part of the Manefay Response Plan, the superintendents and managers were to be available when the mine was at Level Orange and they would meet at the site to decide what to do from there.

Since a Mayday had been declared, the Rio Tinto Emergency Response System dictates that in an emergency, an IC Center is to be set up and led by an incident commander who has control and responsibility of the emergency site, and Collin became the incident commander. The incident commander's job is to take charge of the situation to primarily protect people and secondarily protect property. The incident commander has full authority to use any means or resources required to meet the primary objectives—protect people and property. Once everyone was on-site, the incident commander had many managers, superintendents, supervisors, and operators to call upon to get the work done. Most of the managers, superintendents, and several supervisors went to the mine and initially met at the Administration Building, but the team had problems setting up in that building, so they moved to the Operations Building at the 6190 Complex.

Most of the people that responded to the Mayday had been at the mine since early in the morning, focused on evacuation of the mine, and had left the mine only a few hours earlier. The fact that the Manefay had acted differently than expected, resulting in a Mayday, would test the endurance of the team as they worked to resolve the next set of problems.

At approximately 9:40 p.m., I got the call from Matt Lengerich that the Manefay had failed. My initial response was, "OK. I am on my way to the mine." But Matt countered, "No. You have to stay home and go to bed. Fresh people will be needed in the morning." Matt wanted me to take charge of Incident Command in the morning, so I was going to have to get some sleep. One of the most difficult things that I have ever done was to stay home that night, even though I knew Matt was right about the need for fresh resources at daylight.

At the same time as Incident Command was being established, Matt and the Senior Leadership Team at Kennecott met as a Business Resiliency and Response Team (BRRT) in the board room at the Daybreak offices.

The BRRT had a number of duties. The team's first job was to make sure the incident commander had the required resources. It also had strategic duties such as managing the communications to the press, regulatory agencies, and the larger Rio Tinto Corporation. Stephane Leblanc as chief operating officer of Kennecott was the leader of the BRRT, and he had a wide array of support staff that included Procurement, Communities, Human Resources, Safety, and Operations experts to call upon.

One of the unexpected results of the Manefay slide was that there was not a large release of dust. There had been snow the day before the slide, but up until 9:30 p.m. on the day of the slide, the weather had been dry. Shortly after the slide, a drizzling rain commenced at the 6190 Complex, which changed to snow at higher elevations of the mine. Considering the explosive nature of the failure and the distance the mass traveled, it was surprising there was not a large plume of dust covering the entire area. However, according to an eyewitness, there was no dust as a result of the failure. This is particularly odd, considering that the Manefay mass was supposed to contain little or no water, so dust was expected.

Because of the unexpected explosive nature of the Manefay failure, a decision was made to send the remaining employees working in the Cornerstone area of the mine home for the remainder of the night shift. The immediate task for many of the superintendents and supervisors was to make sure that the evacuation of the rest of the mine was communicated and all employees were accounted for as they left their work areas to travel to their personal vehicles located in the Lark parking lot at the bottom of the mountain outside the mine.

The Bingham Canyon Mine uses a bus system operated by a third party to transport employees from the parking lot to either the 6190 Complex (close to the Manefay failure and now the location of Incident Command) or the 6880 Complex, the location of Cornerstone's offices and facilities. Supervisors were taking a head count of their employees boarding the buses while superintendents checked to make sure all of the Cornerstone working faces had been evacuated.

The Manefay failure had behaved contrary to historical failures. Instead of slowly failing over a period of days or weeks and reaching stability, a larger-than-expected mass had exploded into the pit in a matter of seconds. However, now that the Manefay had failed, the priorities were to continue keeping people safe, determine the effects of the Manefay on the mine, and make a plan to return to production—at least that's what the next steps were supposed to be.

The Second Failure

After the word went out that the Manefay had failed, managers and superintendents rushed to the mine to evaluate the impact and start the process of returning to production. The IC Center in the Operations Building at the 6190 Complex was a hive of activity as the IC Team tracked the progress of the Cornerstone evacuation. There was not only constant communication with superintendents and supervisors accounting for people, but also with the BRRT located at the Daybreak offices. Bill Snarr, one of the operations facilitators, and Anthony Hoffman, a superintendent, were outside the IC Center when they heard rock falling and they began to investigate. Bill and Anthony could hear an almost constant sound of rock fall coming down the steep ridge east of the Bingham Shop, which was approximately 500 feet from the Operations Building. They used a high-powered spotlight to inspect the highwalls without getting too close. They could also see the upper part of the slope and ridgeline because the drill crew that had been evacuated earlier in the evening had not shut down their generators, so lights from the drill illuminated much of the slope.

As Bill and Anthony were looking up at the slope, they saw a large blue-green flash of light, possibly from the substation near the top of the mountain. Shortly after the flash, Bill, Anthony, and two others listened as the rock fall became louder and louder. Then they began to hear and feel a loud rumble in the ground. Bill reported that the rumble felt almost like a ridge of thunder coming toward them. The rock fall and rumble continued to increase until 11:05 p.m., when a loud C-R-A-C-K was heard, and, as they watched, the entire skyline to the east fell away into nothingness as a second major landslide occurred. This was a tremendous shock, because everyone believed that the Manefay slide was over and the critical danger had passed.

According to the seismic data from the College of Mines and Earth Sciences at the University of Utah, the second slide happened as quickly as the first. Within approximately 90 seconds, the second mass fell more than 2,000 feet in vertical distance. The seismic recordings were able to capture the entire duration, which indicated that the failure mass started as solid ground, traveled more than 1½ miles, and came to a full stop in 90 seconds. By my calculation, the average speed was then nearly 60 miles per hour, but the maximum speed of the mass must have been much greater since it started from a standstill and ended at a full stop.

Bill reported that when the mass hit the pit floor, it felt like a large earthquake had shaken the ground and it made a tremendous noise. Although shocked, the group quickly recovered and ran into the IC Center and hollered at everyone to leave the building—immediately! Based on the tone and expression of the people calling for the evacuation, everyone in the IC Center instantly stopped whatever they were doing and headed for the exits. The Operations Building was just 500 feet away from the location of the second slide, and some of the rock-fall material was even closer.

Once again, the Manefay did not act like any highwall failure in Bingham Canyon's history by failing in two separate episodes. In all previous events, the highwall would fail and then stop as it reached the angle of repose. Not anticipating that the Manefay could fail in multiple events and the speed in which the Manefay happened, the IC Center had been set up in a high-risk area!

A Third Event

Those in the IC Center were in the process of moving to the Administration Building when a third event occurred. Elaina Ware was on the phone with Matt and the BRRT when more rock fall could be heard on the phone. The ground was shaking again, and Elania had to quickly end the call. This occurred at 11:16 p.m. and at the time everyone thought it was a third landslide. It was later determined by seismologists at the University of Utah that the third event was actually an earthquake that was the result of having the tremendous weight of the Manefay mass move into to the void of the pit, which relieved a tremendous amount of pressure below the failure area.

The IC Team was in full retreat at that point, and the decision was made to move to the newly constructed Copperfield Shop. The Copperfield Shop was 550 feet north of the Administration Building and was believed to be in a safer location. In reality, the Administration Building and Copperfield Shop were nearly parallel to the Manefay and not much farther from the landslide—but there was no way to determine this in the dark.

Knowing everyone's location was critical during this retreat, because people were scattered throughout the area—James Jung was located at the Administration Building, dispatchers were at the Resource Building, and of course the IC Team was at the Operations Building. Employees were sent to all of the buildings to look for anyone remaining in the 6190 Complex. Everybody was to assemble at the Copperfield Shop so decisions could be made on how to account for everyone and then completely evacuate the area.

Once every employee was accounted for, the decision process began on how to evacuate the 6190 Complex. Using the normal access road going into the 6190 Complex was not an option because it passed the Manefay slide area and was possibly destroyed by the failure. The second option was to use the 4-H access road that came into the back of the 6190 Complex. This road passed relatively close to the Bingham Shop, and it was unknown at the time how far the Manefay slide extended to the access road. The BRRT was able to control a remote camera overlooking the 6190 Complex and could zoom in on various areas. Since it was dark, it appeared that the new wall created by the Manefay might actually extend very close, if not into, the 4-H access road, so that option was eliminated.

The 6190 Complex sits down in a valley with steep highwalls on three sides (with the open area being the mine to the south side). The only other access into the complex was a dozer road that went from the 4-H road down to a ramp on the north side of the Copperfield Shop. The problem with the dozer road was that it was too steep for light vehicles or buses. If that access was going to be used, the group would have to hike up to the ramp where a company bus would pick them up. This route would keep people moving farther away from the Manefay failure area. At nearly a third of a mile the distance was not great, but the group would also have to climb nearly 200 feet in elevation. By now the weather had turned cold and rainy—almost turning to snow, making some of the ground slick and dangerous. These conditions along with the stress of the situation would not make for an easy climb.

It was after midnight by the time the group began its ascent to the top of the ramp. More than 20 people were in the group, which was comprised of a wide range of employees, from hourly operators to senior managers—and a wide range of health and fitness conditions. For some the hike was no problem, but others struggled up the slippery incline. The group encouraged each other while Superintendent Don Mallet and senior coordinator Tony Johnson assisted individuals needing help up the slope. The group worked together until everyone was safely in the bus.

It was almost 2:00 a.m. by the time the members of the IC Team made it off the mountain and to the Daybreak offices where they could be debriefed by the BRRT. During the debriefing, Bill Snarr relayed to Matt Lengerich and Stephane Leblanc what he had seen when the second failure occurred. When asked how far the Manefay failure extended, Bill said the failure ran from the Bingham Shop to the substation at the top of the mountain. The substation was more than 600 feet back from the edge of the highwall. This was much farther than expected and they were not sure whether to believe him. After Bill explained further what he had witnessed, he noticed that both Matt and Stephane turned a little pale as they began to come to terms with what had just transpired—that the failure was much different than anyone had expected.

Establishing a New Incident Command Center

It was a little after 5:30 in the morning when I arrived at the Rio Tinto Regional Center to take over the role as incident commander. The BRRT had been on duty through the entire night and the magnitude of the failure was unknown, with the exception of some verbal reports. Daylight started to shed light on not only the accessibility of the 4-H road, but it also provided the first look at the mine after the failure. The reactions from Matt Lengerich and others in the BRRT were of disbelief when they pointed the camera to the pit bottom. *That* could not be the pit bottom—where was the laydown yard with spare parts and maintenance equipment? Where was the equipment left on the Moly Dome so we could start back into production quickly and efficiently? How could the debris have gone over the 300-foot-high Moly Dome? There were so many questions, but at that moment, the next step was to set up a new location for the IC Center.

The next group of BRRT people was arriving at the board room of the Daybreak offices. That facility was going to be too busy to house both the IC Center and the BRRT. The mine had a secondary Disaster Management and Recovery Center situated in a trailer near the Main Gate, which was more than 4 miles from the 6190 Complex, so the decision was made to set up the IC Center at that site. This proved to be an ideal spot because it was in a safe position away from the Manefay slide area, just past the Main Gate, so everyone had to pass through this gate to get to the mine or the IC Center. It was also convenient to a number of offices and trailers, so small groups of people could find areas to work together to solve problems that may arise.

Setting up the IC Center was quick with the help of Patti Hoggan, the administrative assistant to the general manager; Lori Sudbury, the superintendent of Mine Monitoring and Control; and Tim Juvera, manager of asset management. By 8:00 a.m. of April 11, the new IC Center was up and running with phones, radios, flip charts, and a long conference table. Soon the room was filled with activity from a wide variety of people—electricians, EMTs, geotechnicians, and engineers.

Determining the Priorities

Once the new IC Center was established, the work of determining the extent of the damage and resulting priorities started. For the IC Team, the overarching task was to keep people safe, followed by preventing additional damage to equipment. The initial priorities included establishing control of access to the site, restoring power to maintain geotechnical monitoring, and assessing the damage that had been done by the Manefay. Later in the day, a new priority came forward, which was to account for the owners of several cars remaining in the employee parking lot from the night before.

Establishing Control

The first step was to establish control of the overall situation and determine other priorities. Given that there had been more than one failure and there was a possibility of others to follow, controlling access to the area became the highest priority. The mine had two primary access points. The first was the Main (Lark) Gate and the second was the employee parking lot. Both of these access points led to a single paved road that went up to the mine. Employees were stationed at the employee parking lot to send people to the Lark Gate. Security personnel at the Lark Gate either turned nonessential people away or sent them to the IC Center. A second checkpoint was set up farther down the road to allow cleared people to travel to the mine. Strict rules for accessing the mine were put into place that included the following:

1. Anyone going to the mine had to have a specific reason or purpose that was approved by the incident commander.
2. Everyone going to the mine had to go through a briefing at the IC Center regarding evacuation procedures. They also had to have a plan as to where they were going, what they were going to do, and how long it would take.
3. Every person going to the mine had to have their own radio. If three people were in a vehicle, there had to be three radios. The Manefay area was being continuously monitored by the Geotechnical Team, and if the team saw unusual movement, then every person could be contacted—even if they were separated from each other.
4. Every group going to the mine area had to check in to the IC Center on a regular basis.

Some believed that the procedures were too rigid since most people going to the mine would not be near the Manefay failure area, but these procedures served to limit exposure until we had a better handle on what the risks really were. It also slowed everything down so we could confidently determine and address priorities.

MSHA Response

Shortly after the rules for accessing the mine were established, a Mine Safety and Health Administration (MSHA) inspector arrived to assess the Manefay damage. Mine management considered MSHA to be a key stakeholder and worked toward open communications and building credibility with the agency. MSHA had been informed of the impending failure weeks before the event, updated when the mine went to Level Orange, and notified shortly after the failure.

After a brief update, the inspector wanted to tour the site, so a safety representative and an engineer were assigned to accompany him. The employees knew the mine and could provide radio communications should there be problem.

The MSHA tour lasted a few hours. During that time, the group traveled to many parts of the mine to see the slide from different perspectives. Of concern was that the group traveled to the top of the head scarp (a 600-foot-long overly steepened wall left as a result of the landslide) to look down on the Manefay failure area. That head scarp was considered a high-risk area since its stability was unknown. A constant challenge with inspections is balancing the need to see what the situation entails with the necessity of keeping the inspectors safe. This issue should be considered in greater detail in large events such as the Manefay.

Based on the inspection, MSHA issued a 103(k) Order on the entire Bingham Canyon Mine. The 103(k) Order refers to the particular section of the Federal Mine Safety & Health Act of 1977, Public Law 91-173, as amended by Public Law 95-164, which states the following:

> In the event of any accident occurring in a coal or other mine, an authorized representative of the Secretary, when present, may issue such orders as he deems appropriate to insure the safety of any person in the coal or other mine, and the operator of such mine shall obtain the approval of such representative, in consultation with appropriate State representatives, when feasible, of any plan to recover any person in such mine or to recover the coal or other mine or return affected areas of such mine to normal.

At this point, MSHA had control of the mine and any work to be done would have to be approved by MSHA. All of the operations had been shut down the night before, so no additional evacuation was necessary to comply with the 103(k) Order. However, MSHA would have to approve plans before any work could proceed as the mine moved forward.

Restoring Power

While MSHA was touring the mine, the next priority was to restore primary power to the mine and refuel the generators powering the communications systems. The communications systems were not only critical for people to communicate with each other, they were also the link for the geotechnical systems to transmit monitoring data to the geotechnical engineers. If the communications systems ran out of fuel, geotechnical hazards could not be detected. A group of electricians, operators, and a geotechnical engineer developed a plan to fuel the equipment. The plan was reviewed with the MSHA inspector when he returned from the inspection before the work began.

KSL News Coverage

Around 9:00 a.m., our plans to get the mine inspected and fuel the generators were interrupted. A superintendent called in to tell us that we needed to turn the television on to the local station, KSL. At first I was reluctant to take time to look at a news story—we had a lot of critical work to focus on, but the superintendent said that the station was showing pictures of the slide and urged us to take a look. We had seen the videos from the remote camera, and had been getting reports that the slide had covered up the pit bottom, but we still did not fully understand what that actually meant, so I reluctantly agreed to have the television turned on.

When the pictures from the news story appeared on the screen, there was total silence in the room. The effect of the Manefay slide was much greater than any of us imagined, and the pictures from the KSL helicopter told a story that would have been almost impossible to comprehend in just words. The slide had not only reached the pit bottom, it completely covered the entire bottom of the pit. All of the preparations and equipment that had been placed in the pit bottom were completely gone, with the exception of a group of haul trucks that had been pushed by the slide into a pile on the far south side of the pit—far from where they had been parked. The pictures also showed the full extent of the area that had failed, the enormous head scarps, as well as the damage to the Bingham Shop. The videos from the remote camera showed many aspects of the Manefay slide, but it was not until we saw the images from KSL that we really understood the magnitude of the damage, the extreme destruction, and the difficult job we faced to get back to normal operation.

The photos and story from KSL made national and international news. Almost instantaneously, people from around the world knew what had happened at the Bingham Canyon Mine. Many who saw those pictures, including employees, thought that there was no way to recover from such massive damage, and that the Manefay sanctioned the end of a mine that had been operating for nearly 107 years.

Everyone in the IC Center was in shock after seeing the pictures. It was an emotional and sobering experience to see our mine near devastation. After a few minutes of looking at the images on the screen, the TV was turned off. The pictures reinforced that we needed to take things slowly and keep everyone safe each and every step of the way. But they also confirmed there was a lot of work to do and we needed to get busy if it was going to get done.

Assessing the Damage

Figure 1.7 • Undamaged Equipment

The third priority for the IC Center on the day after the slide was to determine what equipment had been lost. This was a difficult task because much of the lost equipment was completely buried by the debris. The only way to determine which equipment was lost was to identify the intact machinery and compare that to a list of the machinery that was left in the pit before the slide event. Entering the pit was not an option the day after the slide, and the Bingham Canyon Mine is so large that equipment numbers could not be readily seen. To identify the equipment that had not been damaged, a spotting scope used by hunters was purchased at a local store. With the spotting scope, a majority of the intact heavy mining equipment was identified.

Eighteen of the Komatsu 930SE haul trucks and an Atlas Copco AC DML drill had been parked near the in-pit crusher and were undamaged. The P&H 2800XPC shovel had been moved for preventive maintenance before the mine was evacuated, and a Komatsu WA1200 loader had also been moved away from the ore face. The Manefay debris did not reach the in-pit crusher, conveyor, and tunnel, so they were operational. There was limited mobile equipment, but enough to resume ore production in the mine. Figure 1.7 shows the location of the equipment in the bottom of the pit that survived the Manefay slide. Just as important as equipment, the Keystone access road was still in place, with the exception of a few hundred feet at the bottom of the road that had been covered up by Manefay debris. The Keystone access road could be repaired and reopened within a few days so that people and supplies would have a way into the mine when ore operations were to resume.

Unfortunately, several pieces of large mining equipment as well as parts, supplies, and maintenance equipment stationed in the pit were damaged or destroyed because of the Manefay. Some of the heavy mining equipment damaged or destroyed in the slide included two electric rope shovels, one hydraulic shovel, three drills, several graders, some dozers, and 13 Komatsu 930SE haul trucks. Of the 13 haul trucks that were damaged or destroyed, eight of them had been pushed ahead of the debris flow like driftwood on a river and deposited at the south end of the Manefay debris flow. These trucks were moved hundreds of feet by the rushing debris flow. Ultimately, five of these haul trucks were repaired and placed back into service. One of the trucks that returned to service had minimal damage and just needed the battery replaced, three had moderate damage, and one had severe damage but was able to be rebuilt.

Accounting for People

The next priority arose early in the afternoon of April 11 and took priority over everything else when it was observed that there were several cars parked in the employee parking lot that did not belong to people currently working at the IC Center. The BRRT asked if we were sure that everyone had been accounted for when evacuating the mine. Was there any possibility that one of the cars in the parking lot could be from someone still left at the mine? Although we believed everyone had been accounted for, we could not be absolutely positive. Consequently, we logged all of the license plate numbers and contacted the owners to ensure they were not still at the mine. Within a few hours, the owner of every car was contacted. The cars not belonging to the IC Team members belonged to employees who had caught rides with co-workers because they had left their keys at the 6190 Complex during the evacuation. Although it was a time-consuming exercise, it was important to make sure everyone was safe.

Internal Communications

During the day on April 11, a significant communication effort was established to keep employees and others informed about what was happening as a result of the Manefay failure. Four internal communications were sent to all employees at Kennecott. The announcements started a little after 3:00 a.m. when the first internal communication was sent to all employees that the Manefay had failed, as well as an update that operations had been suspended and that there were no injuries. Social media, such as Facebook and Twitter, were used to inform people that e-mails had been sent and to direct people to the company website for updates.

At 9:00 a.m. an e-mail was sent that discussed the fact that the day-shift employees had been sent home (except for those working in Incident Command) and that MSHA was on-site. The employees were given a telephone number to call if they had any questions regarding reporting to work.

At noon, a third note was sent from the chief executive of the Rio Tinto Copper group in London. This e-mail was sent to all employees around the world in the Rio Tinto Copper group informing them of the Manefay slide. The news articles of the Manefay slide had started to appear on the Internet and there were many questions about what was happening. This widespread communication effort decreased speculation and kept the entire Copper group informed.

The last companywide communication of the day was sent at approximately 1:30 in the afternoon. This e-mail reiterated that all employees were safe and the safety of the mine was being assessed remotely before sending people into the mine. All employees were to report to work as usual. The mine employees would report to the Lark parking lot to receive an update from the mine leadership.

The amount and tone of the communications were extremely important at this point in time. Employees were concerned about what had happened. They wondered whether their fellow employees were safe and if they still had a job going forward. Although the multiple communications did not answer all the questions people had at the time, they at least had as much information as was available, and this built on the trust and transparency the company had established before the Manefay failure.

One of the best, and in some ways prophetic, communications just after the Manefay event was an e-mail from Matt Lengerich, the general manager of the mine, to all of his direct reports shortly after noon on April 11. Matt had been up the entire night and had gone home to get a few hours of sleep before reporting back to the IC Center and then to the BRRT. Here is Matt's message:

> I am on my way into the RTRC and I wanted to share a couple of thoughts with you.
>
> I am incredibly proud to be a part of the Bingham Canyon team. Last night's response was nothing short of extraordinary—my sincere and heartfelt thanks.
>
> However, as great as that is—our best is yet to come. It is our response in the coming days, weeks, and months that will truly set us apart. I can't think of a better team to face these challenges with. As we have always shown, in every decision, we will put people first and Keep Each Other Safe.
>
> Let's **rise to the occasion** [emphasis added],
>
> Matt

This e-mail was important for a few reasons. First, it set the tone of appreciation from a senior leader for everyone's efforts the night before. There had already been a lot of work going into the Manefay, and the fact that there were no injuries or fatalities was significant. Having a senior leader show appreciation reinforced that people had done an exceptional job and should feel proud of the result.

Second, Matt set the stage for the incredible amount of work ahead, yet he had confidence in the people going forward. These words certainly rang true in retrospect when looking at the results of the remediation work and accomplishments that were to come over the next several months.

Third, Matt drove home the point that people and safety were the first priorities. He noted that we need to keep people first—a reminder that we could not let up with our focus on safety. Again, when looking at the results, this may have been a self-fulfilling destiny.

It is Matt's last sentence that became the rallying cry of the company, and consequently the book's title. In just a few simple words—*Rise to the Occasion*—Matt challenged us to do more, be better, and to not take shortcuts.

These few words became the Bingham Canyon motto that people at all levels took to heart. These words were indeed prophetic—because the men and women at Bingham Canyon did Rise to the Occasion when it came to remediating the Manefay.

Family Support

Earlier in the afternoon, my wife Linda had called. She was on a plane leaving Phoenix for Salt Lake City for our week-long house-hunting trip. When she asked me how things were going, she knew there were serious problems when I responded with only a half-hearted laugh. Linda quickly asked if she should cancel her flight and come up some other week. After a brief discussion, we decided that Linda should still fly to Salt Lake City. She knew that I would not have any time to look for housing with her that week—but she also knew that she might be able to help me while she was in town. Over the next week, Linda did look for a place for us to live, and she also made precooked meals that I could just pop into the microwave when I got home from a long day's work.

I think Linda's attitude and support were very indicative of most of the employees' spouses. There was a huge amount of work ahead of us, and without the support of our family and loved ones, the job would have been twice as difficult. Just as most of the employees went above and beyond during and after the Manefay, so did their spouses and families.

Closing the Incident Command Center

As the day was winding down, all of the critical tasks were completed. We had established control for entry into the mine, backup generators were fueled, power was restored so the geotechnical monitoring systems would continue to work, an inventory of equipment had been taken, and all personnel were accounted for. Matt Lengerich relieved me as incident commander at 7:00 in the evening. Most of the team had already been sent home, and since it was getting close to dark, we had stopped all activity in the mine area.

After Matt arrived, we talked for a while about what had happened during the day and the work remaining ahead. It was hard to believe that it had been less than 24 hours since the Manefay had failed—it seemed like so much had happened since then. After a bit more discussion, Matt decided to officially shut down the IC Center. There was nothing more that could be done that night, and the next day the *real* work would begin—recovering from the Manefay failure.

Matt also arranged to have a driver meet Linda at the airport so I would not have to drive after the long day in the IC Center. I did not expect that, but it was much appreciated. I believe it demonstrated the company's true concern for making sure that people stayed safe.

Manefay Dynamics

We had not anticipated the explosive nature of the Manefay slide. Previous highwall failures in the mine were marked by falling rock and took hours or days to completely fail. The Manefay failed in two large events, nearly 90 minutes apart, and each taking approximately 90 seconds, followed by an earthquake 12 minutes later according to the seismic data from the College of Mines and Earth Sciences at the University of Utah. The failure was amazingly fast, and the material flowed more like a snow avalanche than a rockslide. Because of the speed and the avalanche-like way the Manefay failed, a much larger area was affected than the predicted estimates. Figure 1.8 was taken above the Manefay and shows the path of the once solid rock that flowed into the mine.

Figure 1.8 • Avalanche Nature of the Manefay

Figure 1.9 shows the size comparison between the area that the Manefay was expected to affect and the actual impact area. The area outlined in red was the largest expected area of impact before the Manefay, with the green line indicating the minimum area. The blue line is the actual impact area from the Manefay. The actual slide covered more than twice the expected area as well as the entire bottom of the pit. Fortunately, the crusher, conveyor, and tunnel that transports ore from the mine to the Copperton Concentrator were not affected. However, the debris field was within 525 feet of the tunnel—much closer than anticipated. If the debris had covered the crusher, conveyor, or tunnel, ore production at the Bingham Canyon Mine would have been interrupted for several months instead of a few weeks.

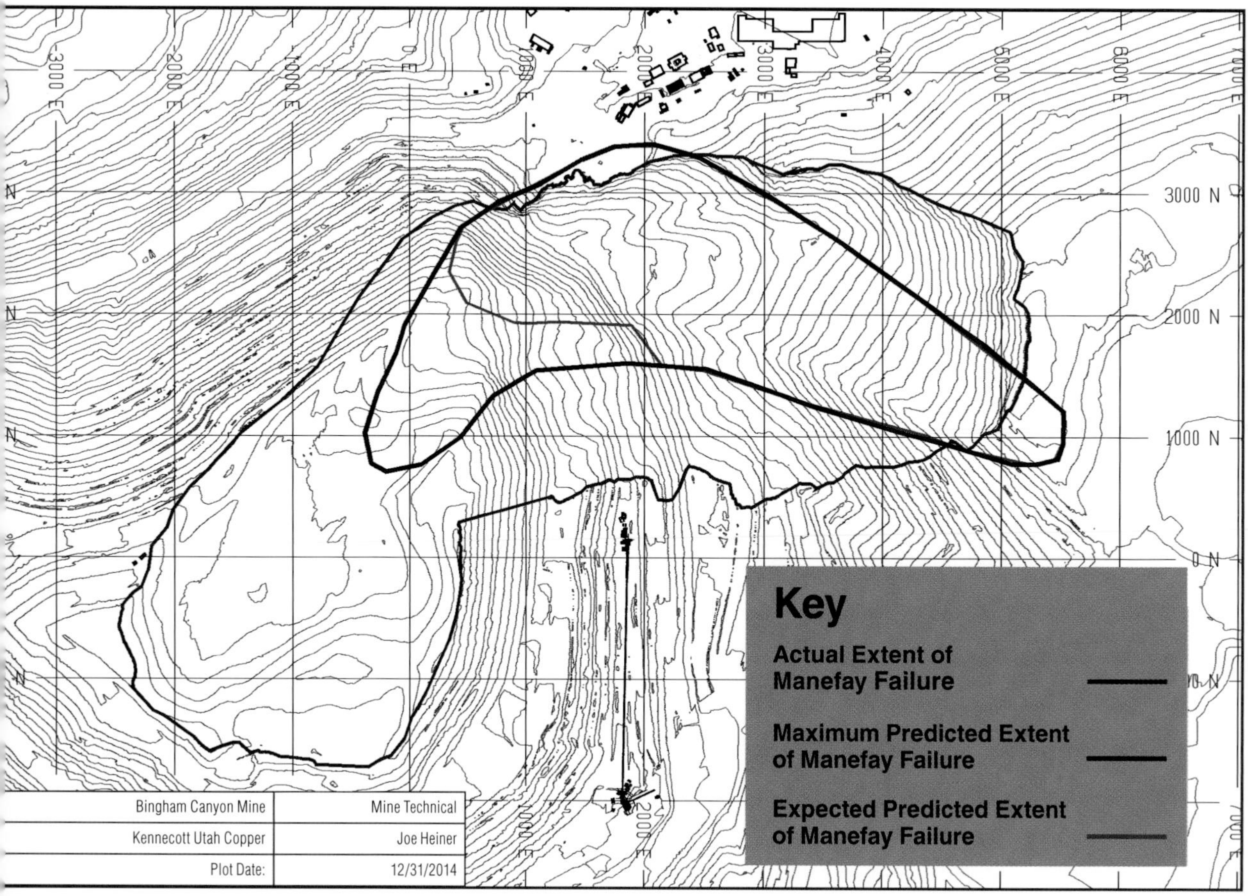

Figure 1.9 • Projected and Actual Impact Area

Instead of falling straight down the highwall to the bottom of the pit as gravity would dictate, the first failed mass began to slide with such great velocity along the Manefay bed that it became airborne as it hit the adjoining highwall. The mass slammed into the west wall, ricocheted off the wall, then swirled around the pit, flooding the bottom of the pit, and pushed over the top of the Moly Dome toward the south wall of the pit. Debris left a "high water mark" (Figure 1.10) over a half mile long and deposited material at least 300 feet higher than the top of the Moly Dome. The debris from the first failure was primarily in-place rock, which is the white or light-colored material seen on the outer edge of the debris field. The second failed mass followed the same path and direction as the first failed mass. Much of the second slide event was old waste material that had been mined decades earlier. This material had oxidized and turned orange or reddish in color. Thus the top layer of the slide debris is orange. The various geologic layers were spread along the length of the debris flow instead of being jumbled into a single homogeneous mass. The result was a layering and swirling of the colors and textures that stretched from the top of the head scarp to the bottom of the pit. It was almost as if the Manefay had become nature's three-dimensional canvass that stretched 1½ miles in distance and nearly 3,000 feet in elevation.

Figure 1.10 • High-Water Mark

Devastation

The sight of the aftermath was overwhelming. Most of the preparations to return to production quickly and efficiently were now covered by more than 300 feet of Manefay debris. Figure 1.11 shows the pit the day before the slide, and Figure 1.12 exhibits the aftermath. The pictures were taken from approximately the same location on the east side of the pit looking west. The white building and equipment on the lower left side of the photo are the facilities used by the Underground Team to build the drainage galleries to dewater the mine. The Moly Dome is the raised flat area in the middle of Figure 1.11 with the pit bottom located on the lower right side. The Manefay failure came from the right side in Figure 1.12.

Figure 1.11 • Pit Bottom Before the Manefay Failed

Figure 1.12 • Pit Bottom After the Manefay

The debris covered almost everything in the pit bottom. Most equipment, parts, and supplies in the bottom of the pit were either completely buried and destroyed or severely damaged by the debris. Figure 1.13 shows the first piece of equipment that was pulled from the Manefay debris. This rubber tire dozer was caught on the edge of the slide. Although the cab was not totally engulfed by the debris, the rest of dozer was extensively mangled.

Figure 1.13 • Komatsu Rubber Tire Dozer

The only remaining mobile equipment that had been stationed on the Moly Dome was now a pile of eight 320-ton Komatsu haul trucks shoved to the far end of the debris flow (Figure 1.14). Figure 1.15 shows truck 495 after it had been removed from the debris. Parts of the truck that were engulfed in the debris were mangled almost beyond recognition.

Figure 1.14 • Haul Trucks in Manefay Debris

Figure 1.15 • Haul Truck Damage

Higher up in the pit, the boom of a P&H 4100C shovel (that the superintendent considered moving) could be seen sticking out of the edge of the debris, as shown in Figure 1.16. The right photo shows a different perspective, with the back of the Bingham Shop at the top. Although the boom looks intact in the photos, some of the metal on the boom is actually remnants of a drill that was also caught up in the debris flow and embedded into the body of the shovel. The debris flow also shoved the shovel, which weighs more than 1,500 tons, much closer to the edge of the highwall.

Figure 1.16 • Boom of P&H 4100C Shovel from Two Perspectives

Figure 1.17 • P&H 4100C Shovel Damage

The force of the debris flow was incredible. Any equipment or parts of equipment that became engulfed in the surge were destroyed. Figure 1.17 shows the P&H 4100C shovel after the debris material was removed. The housing of the shovel and machine deck are almost unrecognizable. Only the heaviest metal components, such as the large gear boxes, were left in place. Even some of the heaviest steel structures, such as the counterweight on the back side of the shovel, were bent and warped.

Later, there would be discussions with the insurance company adjusters about whether some of the equipment could be repaired. They had seen some of the pictures, but even then it was difficult to understand the magnitude of the damage. A month or so after the Manefay, the adjusters were given a tour of the mine and saw first-hand the degree of damage the equipment sustained from the Manefay. After seeing the magnitude of the damage to the equipment on this tour, the adjusters changed their perception of what could be repaired and what was a total loss.

Moly Dome

Before the Manefay slide, the mine had built a maintenance and laydown area on top of the Moly Dome with the belief that the debris flow would not reach the pit bottom—and even if it did, it would not go over the top of the Moly Dome. Figure 1.18 shows the bottom of the pit just days before the slide. The Bingham Shop can be seen on the rim of the pit in the upper center part of the photo, and the Manefay mass is to the right of the shop. In the lower left of Figure 1.18, the Moly Dome can be seen with the maintenance and supply facilities on the flattened area. Equipment can also be seen working well below the Moly Dome.

It was anticipated that the 10% haul road, the primary access road into the mine and the only access road for heavy equipment in and out of the mine, would be damaged in the impending slide. After the slide, access to the pit would be limited to the narrow and winding Keystone access road. Equipment, spare parts, and supplies had been stationed on the Moly Dome so that ore production could resume as quickly as possible after the Manefay slide. The Moly Dome was the only location in the pit that contained an adequate flat area for the maintenance pad and parts laydown yard.

Figure 1.18 • Moly Dome, Before the Manefay

The Manefay did not act as expected and the entire maintenance and laydown area was covered with more than 300 feet of debris. The debris flow not only buried the facilities—the fuel station, maintenance pad, and sump for the wash bay—it buried all of the spare parts and support equipment that had been brought down into the pit so that maintenance could be performed on the equipment that was left in the pit. The parts and support equipment loss was significant. The mine was prepared to maintain the equipment in the bottom of the pit for several months and had stationed enough larger parts and supplies (such as tires) that would be difficult to get down the Keystone access road to keep the equipment operating for up to six months. Some of the more expensive parts and support equipment lost at the bottom of the pit included more than 75 heavy haul truck tires, light field service trucks, forklifts, spare components for electric shovels, and electrical substations for shovels or drills.

The loss of the equipment, parts, and supplies was a blow to the company. Even though two shovels were removed from the bottom of the pit before the failure, the mine took an unknown risk by leaving equipment in the pit prior to the Manefay. One of the options had been to take all of the equipment out of the pit bottom before the failure. Such a move would have protected the equipment, but the mine would have been out of production through August of 2013 when trucks and one hydraulic shovel could have been constructed in the bottom of the mine. Even though much of the equipment was lost, the mine was able to continue to operate with the fleet of equipment that remained.

Shop Damage

The slide damaged more than the equipment, parts, and supplies at the bottom of the pit. The back portion of the Bingham Shop was destroyed and what was left was sitting precariously on the edge of a scarp, as shown in Figures 1.19 and 1.20. The Bingham shop was not salvageable and would have to be demolished.

Figure 1.19 • Bingham Shop Front (left) and Back (right)

Figure 1.20 • Bingham Shop Before Slide (left) and After (right)

In addition to the facilities and mobile equipment, the Manefay did significant damage to the Bingham Canyon Mine itself that would have to be repaired to return the mine to full operation. Looking at the top of the area that failed in Figure 1.21, the slide created scarps (cliffs formed by faulting or erosion) nearly 1.2 miles long that ranged in height from 200 feet to nearly 600 feet. These scarps would have to be stabilized. The Manefay also filled in sections of more than 70 safety benches. Safety benches are critical protection devices that prevent rocks from rolling down the highwalls where they can cause injury or death to people below. To protect the lower benches, these safety benches would have to be cleaned. About 0.6 miles of the 10% haul road was either destroyed or covered with debris and would have to be reestablished.

Figure 1.21 • Pit Damage

Knobs and Scarps

Another remnant of the Manefay was an overly steep piece of rock left in place just above the 6190 Complex that became known as the 6190 knob. Not only was the 6190 knob a steep scarp over the failure zone, it also had a relatively large failure zone on the back side that threatened the contractor's drill and parts of the 6190 Complex. Figure 1.22 shows the proximity of the 6190 knob to the 6190 Complex.

Figure 1.22 • 6190 Knob Overview (left) and Close-up (right)

It is not known why or when the secondary failure on the 6190 knob occurred. It could have been in conjunction with the initial Manefay landslide or the second. One theory is that it was a result of the second landslide running into the wall and basically creating the failure on the other side of the 6190 knob. It is also possible that it was a consequence of the rebound effect that created the earthquakes after the second failure. Regardless, it was probably the rubble from this slide area that the IC Team heard after the second slide which created concern that additional slides were possible and the 6190 Complex should be evacuated as soon as possible.

Besides the immediate concern of the IC Team, the slide on the back side of the 6190 knob created questions about the stability of the knob itself and the safety of the 6190 Complex. The 6190 knob would have to be remediated to get the mine back to normal operation.

Approximately half of the mine's exposed ore supply was covered or encumbered from being mined by debris material. This was a critical issue, because even if ore mining could resume, then ore availability was limited and would run out toward the end of 2013. A majority of the 144 million tons that failed ended up in the bottom of the pit. Ultimately this material would have to be mined and moved out of the pit. Unfortunately, all of this material was waste and not ore that could be processed to recover copper.

The Manefay slide was the largest mining landslide in history. The damage to the Bingham Canyon Mine was immense and the amount of equipment that was destroyed was phenomenal. Despite its gigantic size, amount of destruction, and events that could have been more disastrous, no one was hurt or killed as a result of the Manefay.

A timeline of the major Manefay events is included in Appendix B.

Lessons Learned When the Manefay Failed

The day of the Manefay, as well as the day after, was a roller-coaster ride of high-intensity events followed by relief that there were no injuries or deaths—which was then followed by another event. Many lessons can be learned by understanding how the day of the Manefay unfolded as well how Kennecott responded immediately after the failure.

Some of these lessons, such as having and following a plan, saved many lives. This was a very positive result of the Manefay that can be used by individuals and organizations as an example of how to do things right. Something that we did not do well with was to anticipate multiple events; this is a critical lesson that anyone planning for or responding to an emergency should take into account.

Following are the lessons learned when the Manefay failed.

Have a plan. On the morning of April 10, the operations manager followed the Manefay Response Plan and increased the response to Level Orange. This change in the response level initiated the evacuation of the bottom of pit operations and is the basic reason that there were no injuries or loss of life as a result of the Manefay. Having a response plan in place meant that the only decision to be made was to move to the next level. At that point, the evacuation of the pit was automatic.

Keep communications open. As shown by timely and efficient evacuation as well as the e-mail from the Joy Global manager, open communications the day before and the day of the Manefay resulted in people knowing what to do and why particular tasks needed to be done. The result of this effective communication was that people followed the plan.

Seek an independent review. The greatest surprise of the Manefay was that it acted differently than any previous highwall failure at the mine. The failure exploded quickly in two separate events instead of slowly over time, resulting in debris covering the entire pit bottom. The planning for the Manefay might have benefited by having independent experts review and be a part of the planning process to challenge the belief that the Manefay would act like all previous failures.

Expect multiple events. While planning for the Manefay, we did not anticipate that the failure would occur in two distinct events. Although this may have occurred because we did not understand how the Manefay would act, the result is that the Incident Command Center was located within 500 feet of the second large landslide that put people in harm's way. The lesson from this is there is often the possibility of unexpected events in crisis situations, and Incident Command Centers should be located well away from incident sites.

Safely perform inspections of hazardous areas. Potentially hazardous areas, such as the edge of the head scarp, were entered when company safety representatives and the Mine Safety and Health Administration inspector toured the mine site the day after the failure. Although it is critical for inspections to be conducted by safety and regulatory personnel after a major incident, going into areas of potential hazard places these people at high risk.

Account for everyone during evacuations. The morning after the Manefay, the Incident Command Team used valuable time confirming that employees who had left vehicles in the company parking lot were indeed evacuated. This demonstrated the need for having proper documentation and sharing it with the Incident Command Team in an evacuation scenario.

Communicate with families. Although an event such as the Manefay has a direct impact on the employees and vendors that work in the affected area, the responses and attitudes of family members have a tremendous effect on the employees as well. Communicating to family members is critical for gaining their trust and support.

Keep your keys with you. One of the issues the morning after the Manefay was people did not have their keys for their vehicles to drive home the night of the Manefay. This not only caused problems for the people without their keys, it was an issue for the Incident Command Team.

Before the Manefay

Chapter 2

History of Bingham Canyon Mine

What started out as a discovery of lead in the Oquirrh Mountains south and west of Salt Lake City, Utah, by soldiers from Fort Douglas in 1863 quickly became a series of underground gold, silver, and lead mines. In 1893, Daniel Jackling and Robert Gemmell (a metallurgical engineer and a mining engineer, respectively) studied the mining district and developed an innovative method of mining the ore bodies. Instead of using underground mining methods that relied on producing relatively small quantities of high-grade ore (rock rich in minerals such as copper, silver, gold, and lead), Jackling and Gemmell recommended a large-scale open pit mining method using steam shovels that would mine large amounts of lower grade ore. Their belief was that the increased production rates and greater efficiency would more than pay for all of the extra material that had to be mined and processed.

It took 10 years but in 1903, Jackling and Gemmell formed the Utah Copper Company to consolidate the many small claims and underground mines into one large operation to prove their theory about large-scale surface mining. By 1906, the first steam shovels were delivered to start the mighty Bingham Canyon Mine.

Over the next 107 years, the Bingham Canyon Mine continued to grow and expand. What had once been a mountain slowly turned into a circular pit. By April of 2013, the pit was more than ¾ mile deep and nearly 2½ miles wide—the largest in the world. Figure 2.1 shows an aerial photograph taken of the mine in 2012.

Being inside the Bingham Canyon Mine is both awe inspiring and humbling. At first sight, the sheer size of the mine can take your breath away and you marvel at what humans are capable of building. At the same time, it can make you feel small and insignificant. Distances and scales are deceiving in a mine as large as the Bingham Canyon Mine. When standing on the top edge of the mine and looking down at the equipment working at the bottom of the pit, large trucks and shovels seem the size of toys. But when you get up close, the haul trucks are the size of a two-story house and the tires are twice as tall as a person. The shovels are also massive, and the buckets of the largest shovels can hold multiple pickup trucks. Even the road that goes from the company's office complex at the rim of the mine to the bottom of the pit is nearly 4.2 miles long and approximately 120 feet wide. Yet it looks like a small trail when gazing from one side of the pit to the other.

Like most large mines, the Bingham Canyon Mine was not originally planned to be as large as it eventually became. Over time, the mining and processing operations became more efficient, which allowed it to expand and be mined deeper. As a result of this process, the Bingham Canyon Mine has been expanded in a series of "pushbacks," where stripping from the surface would commence several years before additional ore would be uncovered lower in the pit.

The Manefay slide occurred in the "Giant Leap" pushback in the northeast corner of the mine. This pushback was started in early 2005. Because the ore was approximately 2,000 feet below the surface of the mine, the Giant Leap pushback required nearly five years of overburden stripping before additional ore could be uncovered. Two-dimensional geotechnical modeling and analysis were undertaken based on the mine's rock and geologic conditions as well as an extensive record of previous highwall failures. Based on these analyses, the highwall angles of the Giant Leap pushback were increased to a factor of safety that was historically acceptable—based on the typical failures of the past.

The Manefay was so named by the mine because the slide occurred along a weak zone in a geologic bed called the Manefay. The Manefay bed had undergone significant bending and movement from intrusion of the Bingham Stock, which created an area of gouge, or clay material, that the landslide moved along.

Figure 2.1 • Aerial View in 2012

Finding the Manefay

Given the massive size and complexity of the Bingham Canyon Mine, there is a wide variety of geologic conditions and rock types. Some of these conditions include faults, intrusions, and bedding planes that have made the rock or structures weak and unstable in areas. Over the decades, the mine has experienced hundreds of highwall failures, estimated to be anywhere from a few thousand tons up to a few hundred thousand tons. Figure 2.2 shows a multi-bench failure that involves six benches in the south wall of the mine.

Figure 2.2 • Multi-Bench Highwall Failure

Most failures in the past were relatively small, affecting a single bench over a limited area. As the mine matured and continued to grow wider and deeper, the highwalls became taller and the chances for larger highwall failures increased accordingly, and some of the slides were more than a million tons.

The mine's operating and technical staff recognized that highwall failures were a significant risk to people, equipment, and production. Consequently, they included highwall failures as the number one risk in multiple risk assessments. To help manage this risk, the mine invested in geotechnical expertise by hiring geotechnical engineers and consultants to study and understand how and why the failures occurred. They also invested in state-of-the-art monitoring equipment that became more and more sophisticated over time. This equipment not only helped the geotechnical engineers understand why the highwalls failed, but it provided the information for them to predict that a failure would occur based on relatively small movements in the highwall.

By 2012, Bingham Canyon Mine had all of the critical components in place to identify a large highwall failure before it happened. It had a variety of monitoring systems, ranging from inspections by people to highly accurate radar systems that continuously detected movement in the highwall. At least 10 methods or layers of protection were implemented before the Manefay to monitor the highwalls and keep people safe. More in-depth descriptions of these monitoring systems are provided in Appendix A and include the following:

1. Geotechnical Hazard Training
2. Routine, Documented Inspections
3. Prism Network
4. Extensometers
5. Time Domain Reflectometry
6. IBIS Slope Stability Radar
7. GroundProbe Radar
8. Microseismic Monitoring
9. Geographic Information System Data Display
10. Piezometers

The mine also had a geotechnical engineering team that was dedicated to the protection of the people and equipment by monitoring unusual movements that indicated instabilities and potential failures. The mine employed a consulting firm that had decades of experience at the mine and understood the characteristics of the rock and geologic structures that had contributed to dozens of failures in the past. Armed with this understanding, the geotechnical firm could identify which areas of the mine had the greatest likelihood of having a highwall failure and advised how the mine plans could be modified to reduce the risk of having a large-scale highwall failure. In the areas of greatest risk, the highwall angles would be reduced so the risk of failure was diminished and any failure would be smaller if it did occur.

Even with all of the people, systems, and consultants, highwall failures can never be totally eliminated. Natural systems such as mine geology have too many unknowns and variables to be completely predictable. In addition, the operators may not always mine exactly as planned or dewatering of the pit may occur differently than expected, which could reduce the stability of the pit. For some failures, it is better to let a failure happen instead of trying to do everything to prevent it. People and equipment can be protected because of the monitoring systems, and the cost of preventing the failure can be several times greater than just fixing it afterward.

With each highwall failure, the team's understanding of such failures and the ability to monitor and predict them continued to increase. With three to five highwall failures per year, it can be said that the team's knowledge and ability to predict such collapses was world class, at least for the type of failures that had historically occurred at Bingham Canyon Mine.

The Geotechnical Team monitors the entire mine and identifies areas of the pit that are of particular concern. These areas are then communicated to Mine Operations personnel through tools such as slide hazard maps (Figure 2.3), as an example.

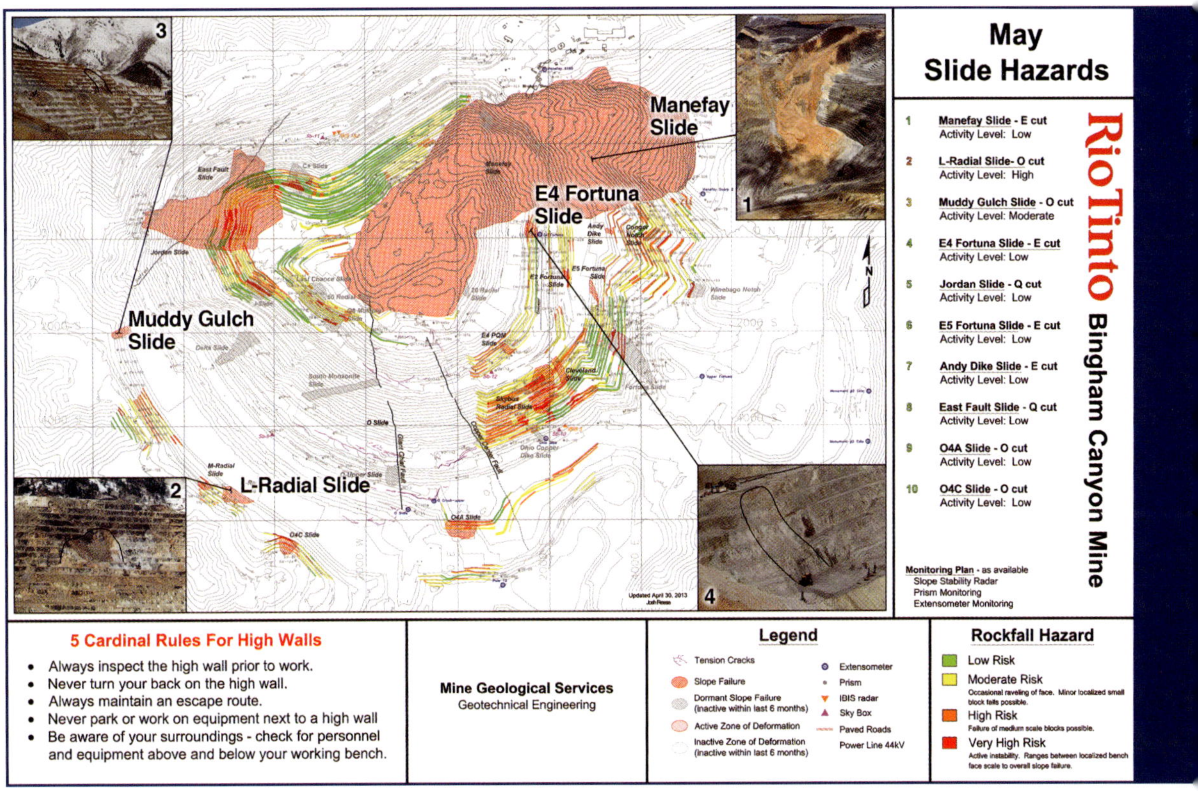

Figure 2.3 • Hazard Recognition Map, May 2013

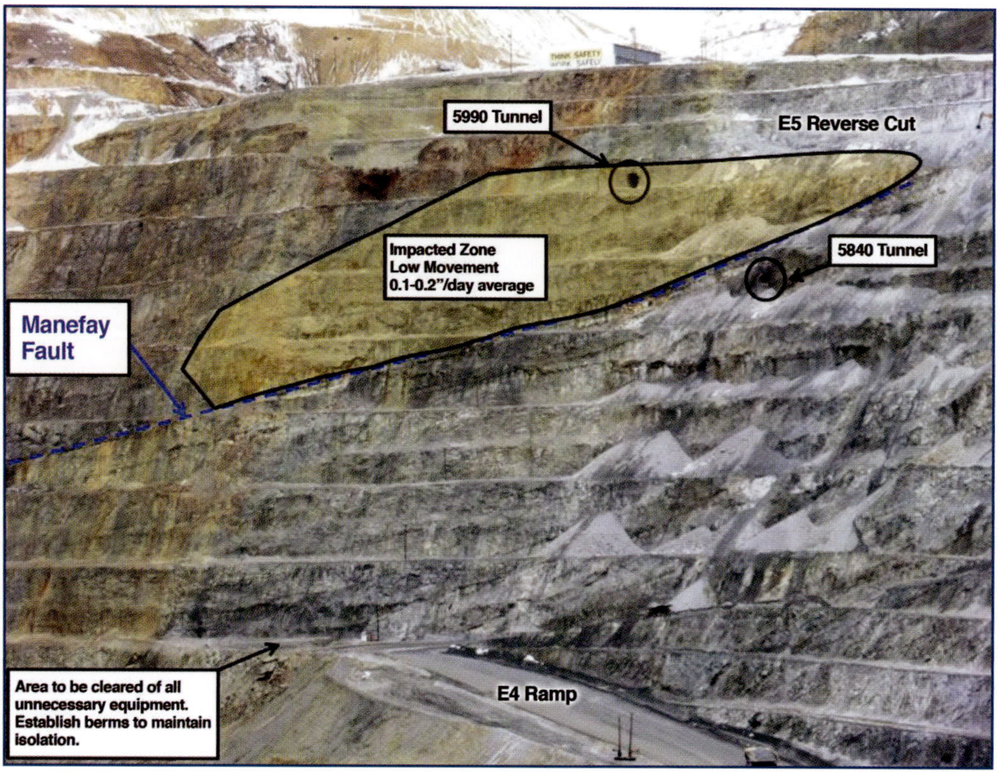

Figure 2.4 • Manefay Movement on January 14, 2013

In June of 2012, the mine experienced a relatively small (150 feet by 150 feet), localized failure on the Manefay bed just above the 10% haul road, which encumbered traffic on the haul road for a period of time. Although the debris had been subsequently cleared, the Manefay was still an area of concern. November of 2012 brought some inclement weather that created some raveling of rock on the former slide area. Because of the raveling and the relationship to the 10% haul road, the Geotechnical Team directed the Operations Team to install a berm around the affected area. This area was placed on the slide hazard map for closer monitoring by the Operations group.

On January 14, 2013, the Geotechnical Department used IBIS radar, a prism monitoring system, and observation of cracks, and consequently identified a block that was localized but below the 10% haul road. The block was believed to be a localized failure that had comparatively low movement and was less than three benches in size, which was indicative of a typical failure on the Manefay. A bulletin was sent out to employees to be aware of the movement and establish isolation berms around the area, as is standard procedure at the mine. Monitoring of the area was increased even further with additional inspections. Figure 2.4 shows the localized instability.

Between January 14 and February 11, 2013, the area of instability continued to grow. Cracks were observed around the Bingham Shop, Visitors Center, and Dispatch Center. On the 11th of February, the IBIS radar showed that the entire wall above the Manefay bed, from the Bingham Shop to where the Manefay outcrops to the east, was moving at a higher rate than the rock mass below the Manefay bed. The Geotechnical Department and outside consulting group were taking measurements to record how much movement was taking place and tried to determine why the Manefay was moving. At 0.10 inch per day, the movement was relatively low compared to previous failures at the Bingham Canyon Mine, but it was still a concern. Figure 2.5 shows how the entire surface of the highwall above the Manefay bed was moving faster than the rock below the bed. The darker the color, the greater the rate of movement.

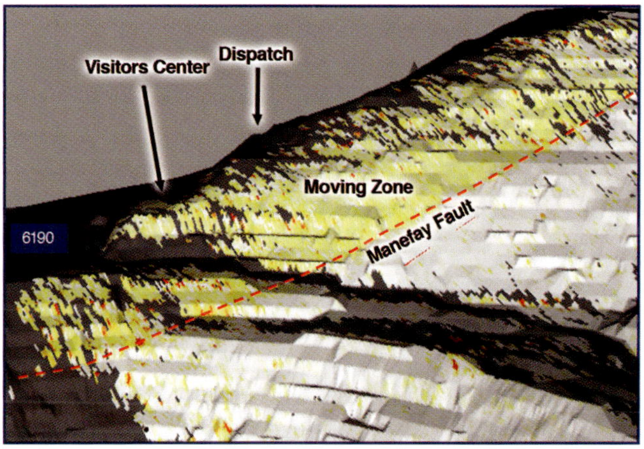

Figure 2.5 • Manefay Movement on February 12, 2013
(Areas in red and yellow have moved between 0.1-0.5"/day)

The Geotechnical Team called on Dr. Zip Zavodni to work with them to help evaluate the data. Zip is a recognized geotechnical expert in the mining industry and was a chief geotechnical advisor for Rio Tinto. Zip was also very familiar with the Bingham Canyon Mine since he had either worked at the mine or supported the geotechnical work for nearly 40 years. Together, Zip and the team gathered as much evidence as possible on the Manefay movement. The Geotechnical Team and outside consultant not only looked at the radar, extensometer, and prism measurements from monitoring systems but also performed field inspections to find and plot cracks and signs of movement. What they found was startling— a large mass above the Manefay bed was not only moving, it was accelerating. Although the movement magnitude was low, the acceleration was of great concern, because this indicated that the Manefay mass was heading to failure.

On February 12, 2013, the Geotechnical Team informed mine management that the mass above the Manefay bed was accelerating, and because the mass was so large, it would probably fail. Whereas most of the previous highwall failures at the mine were measured in tens to hundreds of yards across, this moving mass was closer to a half mile across, meaning that when the Manefay failed, it would be much larger than anything the mine had ever experienced. The Manefay mass would be measured in tens of millions of tons!

On February 14, the Geotechnical Team sent out a bulletin informing the mine about the movement of the mass above the Manefay. In the bulletin they described the monitoring that was being done to track the movement and asked mine personnel to be aware of highwall conditions and report any changes to their supervisor or the geotechnical staff. Figure 2.6 shows the photo and graphics used to show employees the location of the Manefay.

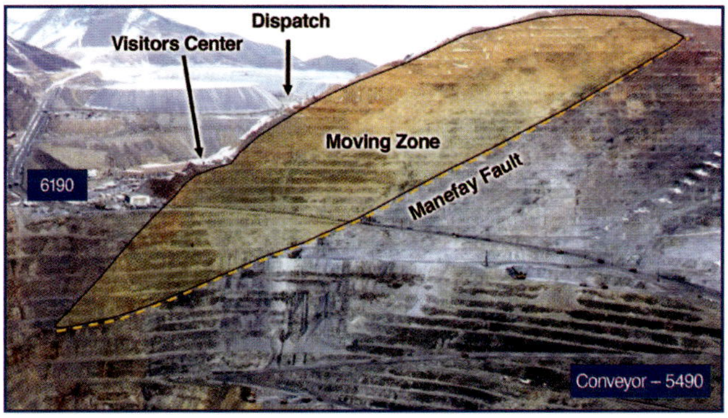

Figure 2.6 • Manefay Bulletin Photo

Even with the news that the failure would eventually happen, there remained a lot of uncertainty. The size of the Manefay was an unknown. Based on the areas that were moving, it would certainly be large, but there was the question of how large. There were also questions about when the Manefay would fail. The movement rate was relatively small at one tenth of an inch per day compared to previous failures that would be measured in inches and even feet per day before failure occurred. So the failure would certainly take many weeks, and potentially several months.

On February 16, Zip Zavodni sent an e-mail to the geotechnical leadership at the mine summarizing many of the discussions with the geotechnical consultant and his thoughts about the Manefay. In the e-mail, he discussed the complex nature of the deformation of the mass and the possibility that there may be an active/passive failure mode. Zip recommended that mining be stopped on the bench below the Manefay mass as well as vacating the Mine Monitoring and Control Building and the Visitors Center. The recommendations also included the development of velocity and inverse velocity charts to track the movement of the mass.

By mid-February, the Geotechnical Team had gathered as much data as possible to answer the unknowns of when and how large the Manefay failure might be. In mid-February, the movement of the Manefay mass remained relatively constant and appeared to have little acceleration, reinforcing the belief that the timing of the Manefay may be later than sooner.

In late February Matt Lengerich, general manager of the mine, invited Martyn Robotham, another geotechnical chief advisor from the Rio Tinto Technology and Information (T&I) group based in Australia, to review and support the geotechnical analysis. Martyn was also very familiar with the Bingham Canyon Mine since he had been the manager of the Geotechnical Department just a few years earlier.

Several activities were started by the first of March. One of the most important was the decision to keep the Mine Safety and Health Administration (MSHA) informed. In most cases, MSHA would be notified after a major incident, but in this case, the company decided to keep them informed well beforehand. The first communication was sent by March 1, and updates were provided on a regular basis from that point forward.

By the beginning of March, the Geotechnical Team had gathered enough data that they started putting together the velocity and inverse velocity charts to help project when the Manefay would ultimately fail. These charts are used to estimate failure time frames but have limited accuracy because of variables such as the type of failure, number of remediation methods, and changing physical conditions (for example, water levels). Figure 2.7 shows an example with data from March 28 to April 10, 2013.

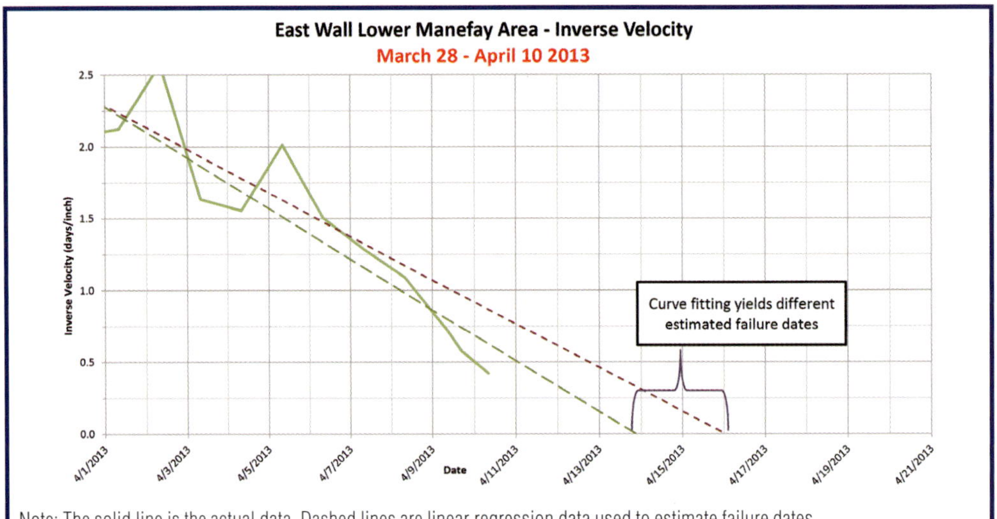

Note: The solid line is the actual data. Dashed lines are linear regression data used to estimate failure dates.

Figure 2.7 • Inverse Velocity Chart

The Geotechnical Department completed the first charts in early March but was hesitant to report the findings to management because of concerns over the accuracy of the findings for the potential failure time frame. Some of the early work showed that progressive failure could happen as soon as late April, which did not fit with their observations of a relative steady movement. That time frame was unexpected, so they continued to gather data to update the graphs until the middle of March to have more certainty in the estimates. By this time, the projected failure date was accelerating and predicted to be mid- to late April.

From late February to the middle of March, the Technical Services Mine Planning Team was working on various plans for the best way to manage the Manefay. The options included unweighting the top of the Manefay mass, leaving a buttress at the toe of the mass, or letting the failure happen and remediating the outcome. The basic premise of the planning was that there would be enough time to make the changes required to prevent the Manefay from failing. At first, this work did not go well because of problems with the mine scheduling models. Consequently, Joan Danninger from the Rio Tinto Strategic Production Planning group was brought in to support the effort and progress was made with the plans.

Strategizing the Manefay

A strategy meeting was scheduled for March 19 with key planning, geotechnical, and management personnel from the mine as well as Zip Zavodni and Martyn Robotham from the Rio Tinto T&I group to determine the best mine plan going forward. It was believed that the decision would be complicated with lots of options and a variety of priorities.

In the meantime, on the 18th of March, the Technical Services Mine Planning Team met to perform a simulation of what would happen if the Manefay failed. Although my team did not know how large or when the failure would actually occur, it was assumed that a large 250-foot scarp would remain and would have to be remediated, which in turn would cause the 10% ramp to be closed for over a year. From this simulation, the team identified the critical infrastructure that was at risk, such as communications systems, power lines, and access into the pit. Other identified critical issues were pit dewatering, potential ore loss, getting people and supplies into the pit, and maintaining ore supply to the concentrator. This simulation would be important because it identified the priorities that needed to be resolved before the Manefay failed.

Strategy Meeting

Aaron Breen, the manager of maintenance projects and an excellent facilitator, was brought in to manage the strategy meeting held on March 19 and to ensure that the best decisions possible were made. The meeting agenda called for an update on current conditions from the manager of the Geotechnical Department, followed by an update on the geotechnical modeling by the geotechnical superintendent. Next, the mine planners would present the mine planning models and decision tree before the entire team worked on which of the options should be pursued further.

The agenda completely broke down after updates were given on the Manefay conditions and geotechnical modeling. The geotechnical manager and superintendent informed the attendees that the most recent movement and modeling indicated that the Manefay would fail sometime between middle and late April 2013. This was a *significant* change from prior expectations of the Manefay time frame; consequently, the team unanimously agreed that the failure was imminent and the only option was to ramp up preparations for the event. This decision not only changed the direction and priorities of the meeting, it also created a new sense of urgency in the entire mine for the next 21 days.

Based on this information, several key decisions were made during the meeting:

- Increase the understanding of the failure mechanism for the Manefay with horizontal and vertical drilling.
- Empty the displays in the Visitors Center and move the building.
- Move forward with plans to facilitate getting people in and out of the pit post-failure.
- Accelerate relocating the Bingham Shop to the Copperfield Shop.
- Build a Trigger Action and Response Plan (TARP) so everyone would know what to do as the Manefay approached failure.
- Put a plan together outlining where to position equipment and supplies to assist in a quick recovery to return to production after the slide.

Matt Lengerich ended the meeting by passing on some thoughts from Kelly Sanders, the president and CEO of Kennecott Utah Copper, where he reminded everyone that "Our corporation values life and safety above all." He also pointed out that Kennecott had the best resources in the world to assist us and that by working together we would be successful. These words were predictive of how the Manefay would be managed.

Changing Gears After March 19

The mid-to-late-April prediction for the failure changed everything. No longer were people concerned with different options about how to prevent the Manefay or if the Manefay was going to fail since it was now clear that the failure was going to occur. With so little time to the estimated date, the primary task was to prepare for the Manefay. The communication of the projected date was also important to the morale and attitude of the management as well as

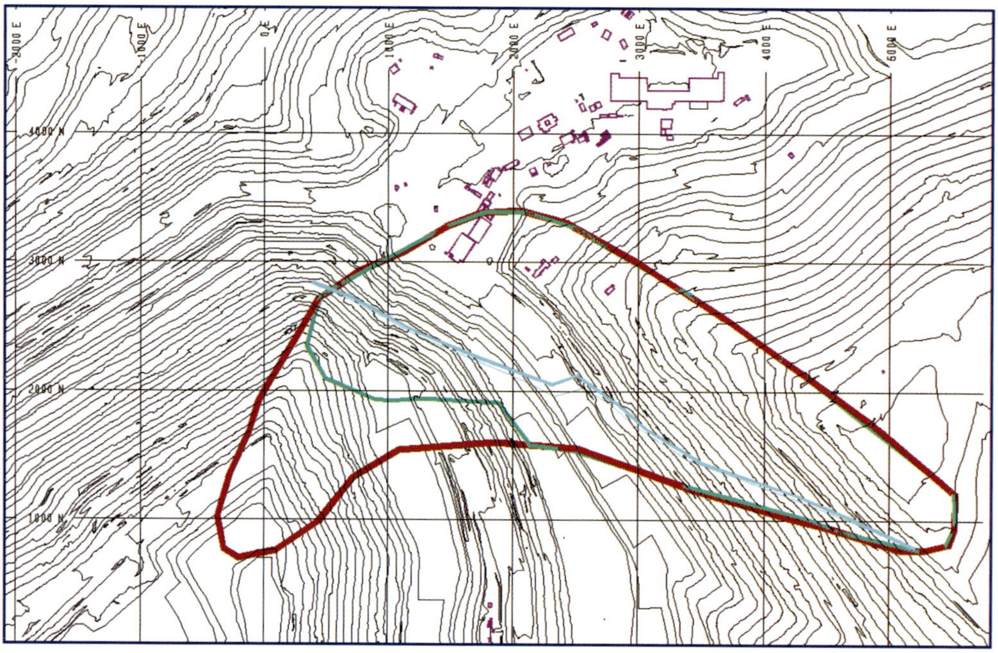

Figure 2.8 • Projected Manefay Failure Footprint

the entire workforce. Even though the time frame was earlier than anticipated, it took away a lot of the uncertainty and stress that had been building over the past several weeks. It also gave more credibility to the Geotechnical Department because it was seen that the geotechnical personnel were being open and transparent. This forthright communication was important throughout the entire Manefay process because it built a much stronger relationship between management and the workforce, which was critical in protecting people and recovering from the slide in a fast and efficient manner.

In addition to when the Manefay would fail, the mine personnel needed to know how large the failure would be and what areas would be affected so people and equipment could be protected. On March 20, Zip Zavodni, Martyn Robotham, and the geotechnical superintendent met to estimate the potential footprint of the Manefay. The superintendent provided an estimated footprint of the slide as shown in Figure 2.8. This estimate included a red line that was considered to be a large failure area for the Manefay, a green line showing the smaller or more likely failure footprint, and a blue line that showed the intact area that was likely to fail. Most of the lines overlapped each other except for the leading edge of the failure.

The geotechnical superintendent was hesitant to publish the information approximating the timing or the shape of the failure. Part of the hesitation was because he knew that any estimate was just that, only an estimate, yet it would be used as if it was gospel. In the case of the failure shape, the outline of the in-place material was based on movements from radar measurements and cracks that were observed and mapped in the field. The runout footprint was based on the assumption that the Manefay would act as a normal rock-fall landslide over a number of hours or days. This seemed to be a reasonable assumption given that the Bingham Canyon Mine had experienced and documented dozens of medium- to large-scale failures over its 107-year life, and that was how slides typically performed. In fact, at least five of these slides had been on the Manefay and they had acted like a rock-fall landslide.

The geotechnical superintendent was correct in assuming that the shape would be taken as fact, because it was used for all the planning and preparation for the Manefay failure from that point forward. Unfortunately, independent experts were not consulted to review or provide feedback of the failure mode, and the geometry of the failure was not challenged outside the Geotechnical group.

The geotechnical engineers were also hesitant to provide estimates about the amount of material that would fail. If the shape was uncertain, then the failure tonnage was even more so. The failure tons were estimated by assuming either a thin wedge-type failure or a circular type failure. Both of these failure modes had happened in the past. Based on the failure shape and cross section of the failure, the Geotechnical Department estimated that the Manefay failure could range anywhere from 60 to 100 million tons.

After receiving the failure shape from the geotechnical engineers, Josh Davis, one of the mining engineers, built a three-dimensional (3D) model of the failure. Josh was one of the young engineers with less than three years out of university and had been recently trained by his superintendent, Cody Sutherlin, to prepare mine designs. Josh modeled the failure as basically going vertical from the cracks on the surface to the bed of the Manefay. Perhaps it was coincidence, but when Josh calculated the failure tons, he came up with an estimate of 120–140 million tons, which was very close to the actual failure tons. Unfortunately, Josh's estimate was discounted because it did not take into account the typical failure modes that the Bingham Canyon Mine had previously experienced. As it turned out, having someone look at the Manefay as a single 3D solid provided better answers than doing the same analysis based on historic failures, mainly because the Manefay was not like the previous failures.

Preparing for the Manefay

The buildup in preparation for the Manefay slide went into full swing immediately after the March 19th strategy meeting, and things started to happen very quickly. Everyone increased their work pace with a renewed sense of urgency. The new work tasks and accomplishments from that point until the failure on April 10th were nothing short of amazing and included moving buildings, constructing a new access into the mine, and reorganizing work locations—all in an effort to keep people safe and quickly bring the mine back into production after the failure.

In many ways, preparation before the Manefay failure was a dress rehearsal for the enormous effort that would be required to return to normal operations after the slide. This groundwork improved the skills of individuals to analyze the work that needed to be done, organize resources required to do the work, and implement effective change. This preparation also developed critical leadership skills of a young and inexperienced professional staff. Some of these skills included the ability to teach, delegate, and rely on others for making judgments and doing the follow-up checks to ensure work was completed.

Preparing for the Manefay failure changed the culture of the company to be much more focused on accomplishing goals while being less bureaucratic. Management and supervisors had to rely on people at all levels of the organization to complete what had to be done for the company to survive. Without this cultural change, the mine would not have been capable of performing at the remarkably high level that was required after the failure.

Communications that the slide was definitely going to happen and that it could occur by the middle of April made preparation efforts the highest priority. Matt Lengerich assigned Lita Madlang, from the Kennecott Community and External Affairs group, to help with communications from the mine to make sure everyone understood this new information. A bulletin went out to the senior leadership throughout Kennecott within a few days, followed by a more personalized letter to all employees a day or two later. In that letter, Matt gave all employees the details that were known regarding the Manefay movement and detailed the actions that were being taken to prepare for the slide. He also asked for everyone's vigilance in watching for dangers and to support the ongoing efforts.

After the March 19th meeting, the previous general manager of the mine was brought in to help direct the movement of buildings, prepare the infrastructure, and establish a method of transporting low-grade stockpile material directly to the live storage piles outside the concentrator. Up until just a year before the Manefay, he had been the general manager for many years until Matt took over. Matt had worked for him at the mine in a few positions, so both men knew each other well. Plus, the former general manager had tremendous knowledge of the mine and its people, their strengths and their weaknesses. Time was short and the work to be done was great, so the two split the responsibilities to accomplish the tasks. The past general manager focused on the infrastructure and low-grade haulage, and Matt took care of everything else.

As the pace of work picked up, the response by people to get the work done was impressive. Most of the staff were already working extra hours, but now they seemed consumed with meeting the challenge as the amount of hours increased, including weekends. Silos and barriers were broken down and people did not follow the normal chain of command. If the general manager needed an answer or problem solved, he went directly to the person that was needed, because waiting a day or even a few hours was too long at the pace things were moving.

Earlier in March, the daily Progress Review Meeting had been changed to meet three times a week, but it was brought back to a daily basis on March 20. Organization of the meeting intensified as well, and a formal Gantt chart was developed to keep track of tasks. Attendance at the meeting also increased with the inclusion of the previous general manager and a variety of other people who were doing critical work that needed to be reported to the team.

Around March 23rd, a decision was made to have the Production Support Department report to the Technical Services Planning manager. This made sense since the two teams needed to work together on several issues. Megan Gaida, who was temporarily acting as the production support manager, was an excellent geotechnical engineer. She had been placed in the production support position to provide her with leadership experience, but she was needed back in the Geotechnical Department to help with monitoring and analyzing the Manefay. Megan benefited from her stint as the production support manager, and just a few months later she was promoted to superintendent of the Geotechnical Department and managed that group through a critical time of rebuilding while supporting the remediation effort.

On the mine planning front, plans were being made for reestablishing the 10% haul road based on the new slide footprint. With the belief of limited impact to the bottom of the pit, the Mine Planning group developed a plan and schedule to resume mining in the shortest possible time after the slide had stopped moving. This included determining the specific shovels, trucks, and support equipment that would be needed to resume ore production in the bottom of the pit. It was important to not only have equipment available to mine, but equipment to remediate potentially dangerous scarps after the failure and reestablish the haul road. It was decided that two shovels would be removed from the bottom of the pit to an area above the 6190 Complex to support the recovery of the haul road after the failure.

Chris West of the Long-Term Planning group started to evaluate other alternatives to gain access into the pit, including constructing a new haul road from the South pushback or Cornerstone cut. Chris had been in the department for many years and had been responsible for much of the design work for the South pushback. As part of his work, Chris wrote up issues that should be considered before the Manefay. Probably the most imperative was a recommendation to completely evacuate the pit before the Manefay failed. This was extremely important because, as part of the options, one possibility was to leave people in equipment during the failure. Chris's recommendation helped to highlight the need for a full evacuation. Unfortunately for the mine, Chris retired less than a month after the Manefay to serve on a mission for his church. Chris's background and knowledge of the pit were a tremendous asset.

Creating a Response Plan

One of the most important outcomes of the strategy meeting on March 19 was the realization that the Manefay was going to happen sooner than later, and the mine needed to respond to rapidly changing conditions. Response plans were required and those plans were communicated to the entire workforce to keep people informed so they could react appropriately as the Manefay approached failure. The response plans would also prepare the mine for recovery from the Manefay so production could resume as quickly as possible.

The first plan created was the Manefay Response Plan, shown in Figure 2.9. The initial plan was put together by the principal advisor for safety and health with the help of the Geotechnical and Operations Departments. The superintendent broke down the plan into five response levels based on a variety of geotechnical triggers. Each level had a corresponding color associated with it so people could quickly see what level the mine was at and if it had changed from earlier reports.

Bingham Canyon Mine – Manefay Response Plan (March 27, 2013 – Rev. 3)		
Trigger levels assessed daily at 7:30 am. Mine Geologic Services will recommend level response to the General Manager or designee. The Level Response recommendation will be communicated daily to the KUC Senior Leadership Team.		
Response Level	**Manefay Characteristic(s) / Trigger***	**Operational Response**
0 (Blue)	**No movement**	**Routine operations and planning**
	Stable conditions	
1 (Green)	**Steady movement with consistent low levels of monitored increase over the past 1–5 days**	**Routine operations**
	Expected failure in the mass	**Operational planning activities for failure**
	Minor to limited deviation in the behaviors of the failure	**Activate BRMP**
		Infrastructure removal from failure zone
2 (Yellow)	**Significant deviation in the monitored behavior of the mass—A confirmed doubling of average Manefay mass movement rates over a 24-hour period**	**Modified mine operations plan**
	Rock fall noted requiring evacuation of current areas and partial closure of the 10%	**Modified mine management plan**
	Active failure expected within days based on significant acceleration in monitored rates	**Restricted access on 10% haul road—spotters and berms in place**
		Evacuation of rock-fall zone. Barricading established.
3 (Orange)	**Active failure expected within hours based on significant acceleration in monitored rates**	**Closure of the 10% haul road, evacuation of failure zone.**
	Active failure, occurring within controlled limits, failure zone is controlled and operational plans are in place	**Temporary stoppage of operations as required**
	Slide behavior requiring complete closure of the 10%	
	Clear and defined deviation in slide mass behavior	**Modified mine operations and asset management plans implemented.**
4 (Red)	**Active failure of Manefay at high/accelerated rate, unexpected causing emergency evacuation**	**Emergency evacuation of the mine and support areas. Operations halted and KUC Senior Leadership Team notified.**
		Incident Command established
		Emergency response planning
*Triggers will be reviewed daily and modified as required by the geotech team.		
Management Team: Lengerich, Ware, Juvera, Ross, Eatherton **BRMP Team Lead:** Leblanc		

Figure 2.9 • Manefay Response Plan

The initial response level 0 (Blue), which meant no movement, was not relevant since the mine was past that point before the Manefay Response Plan was created. The next level was 1 (Green), which indicated that the Manefay was continuing to move but it was not expected to fail for a number of weeks or months. Routine operations would continue, as would preparing for the failure.

The third level was response level 2 (Yellow). This level signified that the movement of the Manefay had progressed to the point where failure was projected to occur within a number of days, up to a week. At this level, the operations would be modified to restrict movement in critical areas around the Manefay failure area and communications would be made to state regulators and MSHA.

Management Response	External / Security Response
Routine activities	Normal
Routine activities	Normal Security Response
Daily management meetings	Notification to MSHA
Maintain normal scheduling and shift rosters	
BRMP	
Activate management team on call within 1 hour response time to the mine	Security—provide information on parking and visiting
Reschedule off-site activities and shift rosters to meet business needs	Regulators/DAQ
Regular updates to the BRMP, increase meeting frequency as required.	Update to MSHA
Increased geotechnical coverage	
Mine management team called to site, but incident command not required due to controlled response	All nonessential personnel asked to leave site
BRMP team placed on standby, daily updates	Regulators/Media
	Update to MSHA
Mine Management team immediately reports to site	Access on authorization of Incident Command only
BRMP activated and relocated to DMR trailer	Regulators/Media
	Interface with MSHA on-site

BRMP = Business Resilience Management Plan; DAQ = Division of Air Quality; DMR = Disaster Management and Recovery; KUC = Kennecott Utah Copper; MSHA = Mine Safety and Health Administration.

The fourth level was response level 3 (Orange). At this stage, the Manefay movement would have accelerated further, and failure would occur within hours to a few days. At this point, operations would be stopped, nonessential personnel would be evacuated from the mine, and the Mine Management Team would be brought to the site. It was believed that Level Orange would be the highest level reached, with the Manefay mass failing in the expected manner. The mine personnel would then wait for the failure movement to finish, at which point the geotechnical engineers would determine when the highwall was stabilized so that the remediation efforts could commence.

The final response level was 4 (Red) and it was hoped that the mine would never reach this intensity. This stage indicated that the Manefay had failed unexpectedly, causing emergency evacuations. At that point, all mine management personnel would report to the mine and the Incident Command Center would be established to start the process of managing a potential disaster instead of a managed event.

The initial Manefay Response Plan was completed by March 25th and went through a few minor iterations. This plan became the basis of additional planning work by the Operations and Maintenance Teams and became the key communications tool for the Mine Management Team.

Bingham Canyon Mine – Manefay – Modified Operations Plan (April 2, 2013 – Rev. 3)		
Response Level	**Date Complete**	**Modified Operations Plan**
0 (Blue)		**Routine operations**
1 (Green)		**Lower pit operations with 2+ monitoring systems active**
		Train at least 6 employees per team as spotters
		Verify equipment capabilities on alternate route
		Work area inspection of alternate route every shift
		Barricade necessary areas to control access
		Evaluate options for effective personnel transportation
2 (Yellow)		**Cease all mining activities in E5**
		Reduce operations in E4 to nonaffected areas only (barricade north end)
		Spotters in place to control access as needed
		Check-in/out system in use at all 3 access locations
		Create additional tie-line(s) in lower pit (E5)
		Activate lower pit equipment strategy (move equipment as needed)
		Test use of alternate access on each team (after vans arrive)
		Berm off areas of potential rock fall
		Positive pass on 10% if required due to width restrictions
		Stock parts staging area with supplies (pipe, pumps, flanges)
		Create in-pit ore stockpile if possible
		Block access on both sides of Bingham Shop (+ berm along backside)
		Clear debris in catch ditch behind substation at 6190
		Begin preparations to stop use of 10%
		Barricade access to top of Manefay
		Beef up berms from Bingham Shop to access road under VC
		Stage vans in lower pit for personnel transport
		Increase lockers in Cornerstone to accommodate 200 additional personnel
		Additional line out room(s) at 6800'
3 (Orange)		**Stop all mining activities (GM approval required for exceptions)**
		Move equipment to 4440' or above
		Ensure all personnel evacuate lower pit safely
		Potential failure zone fully barricaded, including 10%
		All access in and out of pit using alternate route
		Block Mine Access Road at the notch
		Block 6190 Mine Access Road at Sample Building
		Reroute 44kV power line
		Use of traffic control system on alternate route
		Demarcate any loaded blastholes and leave area
4 (Red)		**Incident Command**
		Immediate Evacuation
Management Team: Lengerich, Ware, Juvera, Ross, Eatherton		
BRMP Team Lead: Leblanc		GM = general manager; VC = Visitors Center.

Figure 2.10 • Modified Operations Plan

Shortly after the Manefay Response Plan was developed, Elaina Ware brought together a team to build a Modified Operations Plan based on the same levels and color scheme as the Manefay Response Plan. Figure 2.10 is a copy of the initial modified plan. This gave the Operations Team more detail about what to do at each level, and the Operations superintendents then added additional details for each of the activities. After the initial Modified Operations Plan was developed, Elaina made a critical decision: when the mine reached Level 3 (Orange), operations would be totally evacuated from the bottom of the pit. Both Zip Zavodni and Chris West recommended the change be made, but Elaina made it effective—one of the most critical decisions of the Manefay event.

After the Operations Team completed their Modified Operations Plan, the Maintenance Team put together the Modified Asset Management Plan, also based on the same levels and color scheme. The plan focused on being prepared to perform maintenance on the mining equipment in the bottom of the pit, which was going to be difficult because there would be limited access after the Manefay failure took out the 10% haul road. At the same time, the Maintenance Team had to move from the existing Bingham Shop to the new Copperton Shop, which at this time was not quite completed. Timing was critical, but the Maintenance Team believed they had at least until the middle of April to make the move. Figure 2.11 shows the key tasks for the Maintenance Team for the various response levels.

Bingham Canyon Mine – Manefay – Modified Asset Management Plan
(March 27, 2013 – Rev. 1)

Response Level	Target Date	Date Complete	Modified Asset Management Plan
0 (Blue)	N/A	N/A	Routine operations
1 (Green)	3/26/2013	3/26/2013	Identify and locate staging area
	?		Identify equipment, components, and oils/lube/tanks needing to be stored in staging area
	?		Set up staging area/pad
	?		Move planned equipment, components, etc., to staging area
	?		Identify and perform major component change-outs on shovels, drills, and haul trucks
2 (Yellow)	?		Ensure all planned equipment, components, etc., are in staging area
	?		Move specialty tools to staging area
	?		Test level 3 operations with each team at the same time as operations
3 (Orange)	N/A		All maintenance access in and out of pit using alternate route
	N/A		Maintain equipment in the pit using equipment, components, etc., at staging area
4 (Red)	N/A		All mining activites stop until further notice

Management Team: Lengerich, Ware, Juvera, Ross, Eatherton
BRMP Team Lead: Leblanc — BRMP = Business Resilience Management Plan.

Figure 2.11 • Modified Asset Management Plan

In the early part of March, there had been a great amount of uncertainty for the employees of Bingham Canyon Mine. Everybody had been given communications that the Manefay would fail sometime in the future, but nobody knew quite when. There was a lot of activity as personnel prepared for the Manefay, such as moving buildings, changing offices, and rerouting infrastructure. The workforce also heard that the Manefay failure would be *very large*, and rumors abounded as far as size and magnitude. Some believed that the failure would be much like the many failures of the past—it would be operationally inconvenient, but not necessarily a hazard to people working at the mine. Others believed that the failure would be so large that the Bingham Canyon Mine would not survive and would be shut down permanently.

This uncertainty, combined with the fact that many of the men and women who worked in the Bingham Canyon Mine traveled into a pit that was more than 4,000 feet deep, created an abundance of stress. Since the 19th of March, the Mine Management Team met every morning to discuss what needed to be done and reported on the progress that had been made the previous day. The first question the managers had in this morning meeting was how much the Manefay had moved the day before. This was each manager's guide as to whether the risk of failure was increasing or not. After the Manefay Response Plan was completed, that knowledge also let the managers know whether the response level should be elevated.

Communicating Movement Rates

Shortly after the Manefay Response Plan was developed, an idea was brought up in the morning management meeting: why not share the movement data and response level with all employees every day? By distributing the information, much of the rumors and misinformation could be eliminated. In addition, since all employees would have the most up-to-date information, they would feel more in control of their situation—they could anticipate and be better prepared when the company did move to the next response level.

Matt Lengerich agreed with the recommendation and directed the Geotechnical Team to put together the daily communication. There was some hesitation by the Geotechnical Team, because they did not want to take people away from their monitoring work, as well as concern about how people would react to the information. But by March 28th, the manager of the Geotechnical Team sent out the first daily update, shown in Figure 2.12. This update showed that the movement rate was at 0.30 inch per day and conditions were fairly normal. The Manefay Response Plan had been shared with all employees, so they understood what the Green level meant, but now they had the actual movement data so they could see for themselves how much change there was each day.

Daily Manefay Status Update

Thursday, 28 March 2013

Current Manefay Response Level is GREEN

- Current overall Manefay movement rates are **0.30"/day**
- Current Movement Trend: Steady
  - This is consistent with gradual acceleration of the Manefay moving mass
- Actions Required:
  - Continue to observe and report any changes in slope conditions to your supervisor

Figure 2.12 • Response Level Green

Meetings were held to inform the employees about the Manefay Response Plan. In these meetings, the supervisors and superintendents stressed that if an employee believed a situation was too hazardous, they would not be required to work in that setting. The employees now had the necessary information to determine for themselves whether they felt safe going into the mine—and if they did not, then they did not have to go down into the pit. They were also reminded that any employee who saw an unsafe situation could shut down the operation. If they saw unusual rock falls or cracking in the ground, especially around the Manefay, they were to call their supervisor, Production Control, or the Geotechnical Team to have the area checked out. If they believed it might be an immediate hazard, they not only had the right, but the expectation, to stop all work in that area.

This method of communication with the workforce was highly effective. There were many responses of appreciation for the information that was shared on a daily basis. One year after the Manefay slide, Kennecott held a celebratory breakfast to mark the milestone of the mine recovering from the crisis. There were many stories that were told at that event, but the one thing that came up most often was the daily communication and how it not only brought some certainty to what was going on, but it also increased the workforce's confidence in the Bingham Canyon Mine's Geotechnical Team and overall leadership.

Moving the Assets

Some of the most impressive work before the failure was performed by the Production Support Department. The team was normally responsible for capital justifications and managing consumables, as well as overseeing the Pit Dispatch Team. Lori Sudbury was in charge of dispatch operations and very self-motivated. In addition to her normal duties, she took on the relocation of the Dispatch Center before the March 19th meeting and the pit communications system after the meeting. The communications system was critical and included the fiber optics, cell phone tower, power lines, and Rajant system that connected all of the geotechnical monitoring systems to a central database so the geotechnical engineers could monitor the pit slopes. If these systems had not been moved before the failure, the lack of communication would have significantly delayed the start of the remediation effort. It can be difficult to manage outside contractors to accomplish work quickly, but Lori was able to have the communications system moved before the Manefay failed. Figure 2.13 shows some of the critical infrastructure that had to be moved before the slide.

Figure 2.13 • Cell Phone Tower and Power Lines in Failure Area

The section of the Production Support Team that usually justified capital expenditure was led by Jessica Sutherlin and included some young, motivated, and innovative engineers—Chris Haecker, Karen Bakken, and Sunny Konduru. This team moved buildings, bought vans that would be critical for getting people in and out of the mine, and organized logistics for managing the Keystone access road that would be a potential bottleneck after the Manefay. The team was able to help organize moving the Mine Monitoring and Control (MMC) Building as well as the Visitors Center, bring in five new temporary office trailers, and move employees from high-risk buildings to safer locations. Since the Keystone access had been established, the team did practice runs with different types of vehicles to ensure that critical parts and supplies could be transported to the pit. They also performed time studies to determine the time required to transport people and vehicles to the bottom of the pit so they could set up a process to manage travel up and down the road. This work turned out to be extremely valuable when pit operations started after the Manefay.

Eric Cannon was temporarily assigned to the Production Support Team and was important because of his maintenance experience. The team worked with the Asset Management and Operations Teams to put together a plan to build a pit-bottom maintenance and warehouse center.

Almost every employee paid close attention to the Daily Manefay Status Update as the daily movement rate continued to increase. On April 5, the rate moved up to 0.60 inch per day, which was double the 0.30 inch per day on March 28. Consequently, the decision was made to move to Level Yellow. The Manefay mass that was moving had been broken down into two areas that were moving at different rates, but had now equalized at 0.60 inch per day. Most employees were expecting the mine to move to the next response level, so they were not surprised, even though it indicated that the Manefay may fail sooner than the predicted mid-to-late-April time frame. Yellow meant that the Manefay could fail in days to a week. Figure 2.14 shows the Daily Manefay Status Update that informed everyone that the mine had gone to Level Yellow.

Daily Manefay Status Update

Friday, April 5, 2013

Current Manefay Response Level is YELLOW

- Current 'Upper Manefay' overall movement rates are **0.60"/day**
- Current 'Lower Manefay' overall movement rates are **0.60"/day**
- Current Movement Trend: Gradual acceleration with localized increased movement
- Actions Required:
  - o Implement spotters at key locations
  - o Prioritization of activities in alignment with Yellow response
  - o Continue to observe and report any changes in slope conditions to your supervisor
  - o Limit non-essential travel on the 10%

Figure 2.14 • Response Level Yellow

The daily status showing the change to Level Yellow went out, and supervisors notified all employees so the response plan could be implemented. Although the pace of preparation increased after the geotechnical engineers had informed management that the failure could be as early as mid-April, it now went into hyper-drive. All departments were completely focused on making sure that critical work was done to keep everyone safe during the failure and to prepare for the return to production afterward.

People at all levels of the company stepped up, took on new challenges, broke down barriers between groups, and performed at tremendously high levels during the last few days before the Manefay failure. Decision making was pushed down to extremely low levels in the company, which is a credit to Matt Lengerich. Had he not implemented this change, much of the work would not have been completed before the Manefay failed. For example, it would have normally taken a manager-level person or team to make a decision about where to place the MMC Building after it was moved from the Manefay mass. But with the short time frames, that decision was made by a business analyst and an engineer. Most people were surprised when the building showed up next to the Resource

Building in the 6190 facilities area, yet they were also supportive of the decision and relieved that the building was out of the way of the slide. There could have be a lot of second guessing and tweaking of decisions, but on a whole, the thought process and decisions were excellent, especially considering how many activities were going on at any one time.

Since escalating to Level Yellow, spotters were now stationed to watch the highwall at all times, and traffic in and out of the pit was minimized. Two shovels were trammed out of the pit so the remediation work could begin as soon as possible after the failure was completed and stabilized. Haul trucks were repositioned in various locations of the mine. The newest and most reliable trucks were parked on the Moly Dome in the pit bottom. These trucks had the best availability and would require the least amount of major component exchanges in the coming months.

The Asset Management Team had some of the biggest hurdles to overcome. They had been planning to be moved out of the Bingham Shop by April 15. On April 6, the day after the mine went to Level Yellow, Matt Lengerich gave them a new schedule: have everything out of the Bingham Shop by April 9, when the shop would be closed and locked. At that point, no one would be allowed in, no matter the reason. The Bingham Shop was in the potential disturbance zone of the slide area, so anything left in the shop after the 9th would possibly be lost forever.

The Asset Management Team was faced with another challenge: many of the mechanics stored their personal tools in the Bingham Shop and needed them to be moved to the Copperton Shop, but they also needed the tools to complete the outfitting of the maintenance pad on the Moly Dome. It would be much more difficult to get the parts, equipment, and tools down to the bottom of the pit to keep the production equipment operating after the Manefay cut off the 10% haul road, so preparing the Moly Dome area was a high priority.

The Asset Management Team came together those few days between April 5th and 9th and completed both major tasks. Figure 2.15 shows the inside of the Bingham Shop just before it was locked on the 9th. Just a few days earlier, the shop had been filled with mining equipment in various stages of repair, gigantic jacks and forklifts and other maintenance items to repair the mining equipment, and dozens of large tool boxes used by the mechanics. Now everything was gone and the building sat quietly, waiting for its fate whenever the Manefay decided to fail.

Everyone at the mine was working at a feverish pace during the final countdown to Manefay. The Production Support Team was finishing several critical tasks, such as moving the Visitors Center. Figure 2.16 shows the Visitors Center being prepared to be moved off of the Manefay mass. The Visitors Center was a well-known landmark and a popular tourist destination, as hundreds of thousands of visitors came each year to learn about and look down into the massive Bingham Canyon Mine. The vision of having the Visitors Center caught up in the Manefay was one that no one at the mine wanted. The building was moved by April 6.

Figure 2.15 • Last Photo Taken of the Bingham Shop

Figure 2.16 • Moving the Visitors Center

Besides moving the Visitors Center, the Production Support Team was completeing the installation of four double-wide trailers at the 6190 Complex so people that had offices in the slide zone would have a safe place to work. A temporary warehouse structure was also installed to store parts from the Bingham Shop. Figure 2.17 shows the location of the temporary office trailers, the temporary warehouse, and the MMC Building that was moved to prepare for the impending slide.

Figure 2.17 • 6190 Complex

The Production Support Team also worked closely with the Maintenance Team to prepare the maintenance pad at the bottom of the pit. Although there was not enough time to construct a building, the team was able to build a pad to perform maintenance and to store a 60,000-gallon diesel tank farm, more than 75 spare haul-truck tires, and spare engines for equipment. Basically, any large items that were going to be difficult to move down the Keystone access road were taken to the pit bottom before the failure and stored in what was considered a safe location at the top of the Moly Dome. Figure 2.18 shows the location of the maintenance pad and fuel station in the pit bottom the day of the Manefay.

Figure 2.18 • Pit Bottom Maintenance Pad

Karen Bakken completed traffic control plans for the Keystone access road that was finished by the Operations Team. They put together a plan to manage traffic up and down the road as well as tested the capability of various types of trucks, such as fuel and maintenance trucks, to traverse the road because these vehicles would be critical for supporting the ore operations while the 10% haul road was being repaired.

Are We Ready?

On April 9, the Geotechnical Department was ready to recommend that the mine should go to Level Orange on or before April 11. The most recent inverse velocity chart predicted that the failure would occur between April 14 and 18. During the day, telephone and e-mail conversations between the Bingham Canyon Mine's Geotechnical Department and the principal consultant reviewed the most recent data, and it appeared that shorter range projections indicated an earlier failure date than previous long-term projections. This and the fact that a relatively large crack had developed on the haul road near the Bingham Shop created a lot of unease in the Geotechnical group. By early afternoon on the 9th, the outside consultant recommended going to Level Orange on April 10.

Zip Zavodni, the Rio Tinto geotechnical expert, was not available during the day of April 9. He had requested the most up-to-date movement data, and late in the evening of April 9, he was able to provide insightful commentary. Zip noted that the failure date prediction was more precise as the collapse point approached. However, he believed that we had reached the limit of our ability to refine the failure date prediction further. This, combined with Zip's experience that the Manefay had a history of "sudden rapid acceleration before collapse," lead to his recommendation that the mine proceed to Level Orange.

By the end of the day on April 9, I remember thinking: "We are ready." The buildings had been moved, infrastructure relocated, the bottom of the pit was ready to start remote operations and maintenance, and the workforce was armed with the response plans for evacuation and reacting to the eventual failure. We had plans to continue ore production and remediation of the Manefay slide. There was always going to be more that could be done, but all of the critical work was complete. The mine had come together and did what many considered to be impossible—to be prepared for the largest geotechnical event in mining history.

In my mind, just one more thing had to happen, and that was the Manefay had to fail—and fail quickly—so we could start the remediation work as a part of the process to return to normal operation. My biggest worry was that the Manefay would start to fail but take weeks or months to finish; as long as it was moving and unstable, we could not start the recovery process.

Then April 10th came...

Lessons Learned Before the Manefay

Know the greatest risks. As a part of the Rio Tinto Group, the mine is required to go through both high-level and detailed risk assessments. Part of this process is to identify risks that could pose a significant impact to people or financial well-being, including premature closure of the mine. Because of the number of smaller highwall failures in the past, the complex geology, and the fact that the mine is one of the largest in the world, a large-scale failure that destroyed key infrastructure was the one identified as the greatest risk to mine operations.

Monitor the greatest risks. Because a highwall failure was identified as the greatest risk to both people and the business as a whole, significant resources went into building a world-class geotechnical monitoring program. This program has several levels of protection that range from field inspections to the use of cutting-edge technology that uses radar systems and lasers to measure movements within one hundredth of an inch. It was because of the vigilant monitoring of the highwalls that the Manefay failure was predicted and lives were saved.

Don't wait for perfect data. By the first week in March, the Geotechnical Team had started to develop inverse velocity charts that predicted the failure could happen in middle to late April. Because of uncertainties about the accuracy of the results, the Geotechnical Team was hesitant to communicate those dates and waited until March 19 to inform mine management. It was this communication of the impending time frame that created a sense of urgency and set several actions in motion to prepare for Manefay failure. Further delays could have resulted in the mine not being as well prepared for the ultimate failure.

Wear a "black hat." Wearing a black hat refers to one of the "six thinking hats" in a book of the same name by Edward De Bono (1999) and represents the need to be critical of ideas when problem solving. Operations teams are great at solving problems and finding ways to minimize the risk—it is what they are trained to do. But there is also a need for team members to challenge those solutions and even the need to solve a problem, especially in new or unique situations.

Bring in independent expertise. Basically all of the geotechnical analyses leading up to the Manefay failure were performed by internal geotechnical engineers and the one consulting firm that had performed the geotechnical analyses for the mine for several decades. Although having one group conduct the analyses was efficient in that everyone was familiar with the mine and the analyses done to date, this method did not lend itself to challenging basic assumptions or ideas. Consequently, there was not a good understanding of the failure mechanism or the way the failure would act and cover the entire bottom of the pit and destroy a significant amount of equipment. It is not known if bringing in additional independent experts to challenge the analytical work would have identified more accurately how the failure would have acted, but they may have been able to question some of the basic assumptions.

Simulate the disaster before an event happens. An important exercise conducted before the Manefay failure occurred was the simulation that the Technical Services Team performed with the assumption that the Manefay had failed. In that exercise, several critical infrastructure issues were identified that would be required to support the return of the mine to production. These issues included the communications system and critical power lines. As a result, these systems were removed from the failure area and preparations were made to reestablish them quickly. In hindsight, it may have been beneficial to have done additional simulations with other groups or integrated groups.

Maintain open and transparent communication. Sharing the Manefay movement rates with the entire workforce, the Mine Safety and Health Administration, and the business was critical for not only protecting people and preparing for the Manefay, but it also built up the trust that was important in recovering from the failure.

Trust your people. In preparing for a crisis situation like the Manefay, the company pushed down responsibilities and decision making to the lowest level possible. The accomplishments that were made demonstrates what can be done when people closest to the issue are given the ability to be a part of the decision-making process.

After the Manefay

Chapter 3

April 12th was the start of the Bingham Canyon Mine recovery effort after the Manefay failure. The Incident Command Center had been shut down the night before by Matt Lengerich, the mine's general manager, which sent a strong message that the event was over and now the recovery was to begin. The next seven months would be filled with long hours, hard work, tremendous stress, frustration, and uncertainty that came with trying to recover from the largest mining landslide in history—without having a historical reference to determine what needed to be done or how to do it. But at the same time, this period included remarkable teamwork, support, the feeling of accomplishment, the belief that what we were doing was critically important, and the knowledge that the company leadership trusted us to get the job done. It seemed like almost everyone, at all levels in the company, made it their personal mission to single-handedly save the company. For many it was the hardest period of our careers, but also the most exciting and the time of greatest learning.

This chapter and the next detail the amazing recovery of the Bingham Canyon Mine after the Manefay (Figure 3.1). There were many different work streams progressing simultaneously. Each work stream was dependent on the others, and all were critical to the ultimate success of the recovery. The work streams included the Operations group that was able to quickly return to production and achieve rates that exceeded expectations. Another was the remediation work—a task larger than anything that had been done like it before, and one that required new methods, equipment, and processes. Development of new mine plans and budgets were needed because even though the mine was more than 100 years old, the pit had changed overnight and thus the planning had to start anew. Also included was the purchase and commissioning of a large amount of equipment that was required for the remediation work and to replace equipment that had been destroyed. These and many other work streams were critical to the Bingham Canyon Mine being able to quickly and efficiently return to its former production.

In the midst of all this work, Kennecott Utah Copper underwent significant cultural and structural changes that enabled the recovery as well. Communications and employee involvement continued to build on the trust that was created before the Manefay. Leadership within Kennecott and Rio Tinto modified several policies and procedures to reduce bureaucracy and allow decisions to be made by the people most familiar with the problem. Structural changes were made so that low-grade stockpiles could be hauled directly to the mill instead of taken into the pit to be crushed and conveyed. The smelter started to process concentrates from mines other than the Bingham Canyon Mine. Cutbacks had to be made in the downstream facilities to reflect the lower production coming out of the mine—but it was done in a way that built on the establishment of a highly motivated workforce.

The Manefay brought people together on many levels. Barriers and silos between work groups seemed to almost evaporate. Support was flooding in from various departments within Kennecott as well as a variety of business units throughout Rio Tinto. There were also an untold number of offers of support from external businesses, individuals, community groups, and government agencies. Some were from other mining companies that offered to loan equipment and provide expertise. Others came from vendors that could help us with services such as remote-control technology. Kennecott took advantage of many of these offers. Because of the urgency of the work, I am not sure we always remembered to thank all the people and organizations that offered to help, but the outpouring of support was very much appreciated.

The results of the dedication, innovation, and changes made by the employees, management, and supporting vendors were stunning. The mine was able to return to waste production in just over 2 days after the Manefay failure and to ore production in 17 days. Production levels were much higher than projected for the equipment that was still operational, and the remediation effort resulted in the ore being uncovered to prevent an ore gap to the downstream operations from the mine. Importantly, the changes that were made resulted in Kennecott making a small profit

Figure 3.1 • Before and After the Manefay Failure

for 2013—an incredible achievement considering the damage and destruction brought on by the slide, and the fact that many people thought it was the end of the historic mine.

Setting Impossible Targets

On April 12, the geotechnical engineers were closely watching the scarps and highwalls with their monitoring systems and there had not been any large-scale movement since the second failure on the night of April 10. Power had been restored to a majority of the mine, so there was no longer a need to keep refueling the remote generators for the communications and monitoring systems. The entire mine was under the Mine Safety and Health Administration (MSHA) 103(k) Order, so all production was stopped. Staff personnel from the mine were scattered around a variety of locations since the 6190 Complex was shut down. A majority of the managers had found offices in a trailer that was part of the Underground Project Complex just inside the Lark Gate. A second trailer had been taken over by the Geotechnical Team as well as a few short-term mine planners.

A majority of the Technical Services Team and Production Support Team were located throughout the Rio Tinto Regional Center (RTRC) offices in Daybreak. Cody Sutherlin, the Mine Planning Team superintendent, had worked with the Rio Tinto Technology and Innovation (T&I) group at the RTRC to find empty desks after the mine evacuation on April 10. The Mine Planning Team had set up their computers and were starting the job of planning the Bingham Canyon Mine all over again.

The first thing the extended Mine Management Team did at the mine was meet in one of the Underground Team conference rooms on the morning of the 12th to start planning a path forward. It was called the *extended* Mine Management Team because it included the same people that attended the daily meetings before the Manefay and some who were not managers, such as Lori Sudbury and Jessica Sutherlin. Although not in management, they were effective at getting projects completed quickly and efficiently. The meeting also included Anna Wiley, who had recently taken the position of general manager, business improvement and reliability, for Kennecott.

The first order of business was a status report of the current situation. Details included the facts that we were under the 103(k) Order and could not start operations without MSHA approval, the power had been restored, and the Geotechnical Team did not see any large movement of material. We also discussed where the various staff members were currently situated given that the 6190 Complex could not be used. There was a discussion of what was going to be needed for working areas going forward, and it was then that Anna Wiley made one of the most important suggestions for the Manefay remediation planning effort. Anna suggested that we take over a large section of the RTRC offices so the Mine Planning Team members could work together in a single location. Anna encouraged us to use as much space as needed, because the planning effort was critical for the mine to move forward and this would be some of the most important work in the company for the next several months.

My first reaction to Anna's suggestion was that it would be impossible. There were just too many people to move, and politically it would be too difficult. But the more I thought about the effectiveness of the team, it was obvious that she was right. The idea was written down to be addressed at a later time.

The second issue discussed was the psychological and emotional fallout from the Manefay. The Manefay was a tremendously traumatic event for everyone and left a gamut of emotions that ranged from a sense of loss or grief for the mine to a fear of going back into the mine. The people who received the greatest shock were the ones working at the mine on the night of the failure, especially those who were at the 6190 Complex when it had to be quickly evacuated. There was concern that some employees could develop severe anxieties that are often experienced after highly stressful situations. Kennecott offers an Employee Assistance Program (EAP) that is intended to help employees get through difficult times. Everyone on the Mine Management Team was reminded of the program and directed to make sure all employees understood that they had access to help.

In addition to the normal EAP, Kennecott brought counselors to the site so employees would have easy access to them. In the meeting, all managers were encouraged to visit one of the counselors as soon as possible—if for no other reason than to show other employees that it was acceptable to get help if they were having a hard time processing the recent events. If any employees were having problems, they were not only encouraged to talk with a counselor, they were given time off work if it was needed.

After discussions about the status, work locations, and EAP, the topic of getting back into production was brought up. It was at this point that Matt told us that a decision had been made by Kennecott's senior leadership that our target was to have the Cornerstone area back in production by *the end of the day.* It was almost mid-morning already and the reaction of the Mine Management Team, including me, was—"Are you crazy! Don't you know that we just went through a major event and people are in shock? Plus, we are under the MSHA 103(k) Order, so we could not start operations if we wanted to." Almost everyone believed that this was an impossible goal.

After listening to the ranting and raving of what could not be done, the response from Matt, and supported by Anna, was: "We understand. But the goal is still to be in operation by the end of the day, so what are we going to do about it?" After more back-and-forth dialog, as well as another reminder of what the goal was, the discussion started to change from what could not be done to what was needed to start up the Cornerstone area.

Cornerstone was not affected by the slide. It was safe, and none of the equipment was damaged. So we started to ask questions: What did we need to do to get the 103(k) Order lifted from just the Cornerstone area? How do we inform people of the plan and get them to the work area? How could we make the electrical power more reliable? For each of these issues, we started to make plans or set up teams to make the plans to be back into production by the end of the day.

When talking to many employees later about how they felt when the Manefay happened, they compared it to having their house burn down or losing a family member. In this one meeting on April 12, it seemed as if many of the managers went through the five stages of grief or loss (denial, anger, bargaining, depression, and acceptance) in a short period of time. By the end of the meeting, there was a remarkable turnaround, and I believe it was a critical turning point for the Bingham Canyon Mine. It was in this one meeting that the Mine Management Team went from shock and disbelief that a seemingly impossible target was set, to acceptance of a plan and the determination to get it done.

During the next several months, this pattern of setting difficult targets, experiencing disbelief, making a plan, and reaching for the target happened over and over again. But it all started with the first goal of getting Cornerstone back into production. This one decision of setting a seemingly impossible target changed what we did as a company, how we did it, and more importantly, what we believed we could do.

The decision to set the Cornerstone target and change the direction of the company for the next several months was made by Kennecott's Kelly Sanders, the president and chief executive officer, and Stephane Leblanc, the chief operating officer. It was not accidental that such a difficult target was set that resulted in a significant change in the culture of the company. Both men had made a career of changing company cultures and knew that we needed a goal to rally around. They also knew that Kennecott needed to quickly show success to regain the confidence of

Rio Tinto, the community, the employees, and the rest of industry. Many had written off Kennecott as a total loss, and the only way to change that perception was to get back in business as soon as possible. So with one significant decision, these two men not only changed the culture of the company, they changed how the rest of the world looked at Kennecott as a viable company.

Restarting Cornerstone

After the Mine Management meeting, the critical path item for restarting the Cornerstone operations was to get MSHA to modify the 103(k) Order to allow production to resume in that area. In addition to the Cornerstone area, it was determined that access to the Carr Fork Road on the west wall of the mine was needed to maintain the geotechnical monitoring equipment that was already in place and to add a GroundProbe radar system focused on the Manefay area. This radar system would be redundant, so if one radar system were to go down there would be a backup in place and operating.

A plan had to be submitted to MSHA before they would approve a change to the 103(k) Order and allow mining in the Cornerstone area. This would be the first of more than 20 modifications to the 103(k) Order over the next several months. The principal advisor for safety and health was the primary contact with MSHA and was responsible for acquiring the 103(k) Order modifications. He had a good relationship with the MSHA inspectors, and the mine was seen as being proactive to the health and safety of its employees. The principal advisor and the company had also earned significant credibility given that MSHA had been informed of the impending failure well before it happened and had taken the action of evacuating employees before the failure occurred.

The company needed to demonstrate to MSHA that the health and safety of employees would not be put at risk by restarting the Cornerstone operations and accessing Carr Fork Road. This was done by showing the inspector a map with the location of the Cornerstone operations and Carr Fork Road in relation to the Manefay failure area as well as providing a description of the geotechnical monitoring that was going on 24 hours a day, 7 days a week. On the afternoon of April 12, the MSHA inspector visited the site so he could physically see the relationship between the Manefay and other areas. After the inspector was confident that no miners would be put in jeopardy, he modified the 103(k) Order to allow the resumption of mining the Cornerstone area as well as access to the Carr Fork Road.

As the principal advisor was working with MSHA, the Operations Team was putting together a plan on how to actually start the operations. Since the night of the Manefay failure, employees had been told to meet at the Lark Gate parking lot at the start of their shift so they could be directed to work if they were needed. Up until the afternoon shift of April 12, most of the employees had been sent home unless they were needed for Incident Command or had other critical skills, such as electricians who were needed to restore power to the mine. But this started to change with the afternoon shift on the 12th. Some of the employees were sent home, but others were told to stay so they could start the shift that night. Some issues had to be addressed before work could begin. For example, many of the employees had left their personal protective equipment (PPE) in the change house at the 6190 Complex and now did not have access to that equipment. These employees would need hard hats, safety glasses, and steel-toed shoes to be able to work. Some of the shortcomings were easily resolved by bringing equipment out from the warehouse located a few miles away in South Jordan, but other items, such as prescription safety glasses, were more difficult. Some employees ended up going to retail stores to purchase what was needed. The next day, arrangements were made for a supplier to bring a supply of safety boots to the mine so the employees working on the next shift would have appropriate safety gear.

After the employees were supplied with appropriate PPE, the next step was to do the pre-shift work area inspections before allowing people into the work areas. These inspections were detailed, because no chances would be taken in bringing people back to work after the Manefay if it was not safe. After the work areas were inspected, the equipment went through a detailed inspection as well.

By the time all the preparation and inspections were complete, production in the Cornerstone area finally resumed at approximately 1:00 a.m. on April 13—only an hour later than the target that the Mine Management Team (including myself) thought was impossible the day before.

Although the production started out slowly and it was waste instead of ore, Bingham Canyon Mine was back in production in just over two days after the largest mining landslide in history—this was huge. It changed the perception of many external people and employees that the Bingham Canyon Mine would never operate again. Starting so quickly after the failure made a big statement: "Don't count us out—we will rise to the occasion."

Communications

Communications to employees were continuing on a regular basis. Most were updates on the progress being made, but a new announcement soon became necessary. There were many unsubstantiated rumors in the news and social media, so the employees were reminded that if the company had not supplied the information, then it should not be considered factual. In addition, employees were told that communications from the company to the press should only come through official spokespersons. Some of the news stations had interviewed employees who were not official spokespeople for the company and they did not have all the facts regarding the Manefay.

During the afternoon of the 12th, Kelly Sanders sent a communication to all employees thanking them for keeping each other safe. Kelly also noted that the mine was at a defining moment in its 107-year history and that basically we had a chance to write the history of the mine from that point on by what we did going forward. Kelly challenged everyone to "rise to the occasion" so that we could define our own future. In effect, Kelly told us that we could control what happened from that point. These were empowering statements, and they became the basis for how the work was managed throughout the recovery from the Manefay.

The internal communications continued to be sent to all employees on a regular basis to keep them informed of the progress while recovering from the failure. The Communications Team sent out an e-mail, updated the web site, and posted on social media what was happening at least once a day for 2½ weeks, when the first ore was mined and sent to the crusher. After the first ore was delivered at the end of April, the frequency of the formal communications went to every other day for a week and then extended to once a week. The weekly communications continued, with a few exceptions, until the end of November when the mine was able to uncover the next cut of ore and return somewhat to normal operations.

The Communications Team sent out 46 internal communications regarding Manefay updates during the eight months from April through November. This work by the Communications Team that documented many of the achievements in recovering from the Manefay was a terrific effort and hugely important. Not only did it keep the employees informed, it continued to build on the trust and understanding between management and the entire workforce. This trust and understanding was the foundation for the cultural changes that were taking place to build a stronger and more capable company that would not only result in an amazing recovery from the Manefay, but help the future success of the company.

The communications also gave the downstream management and workforce a better understanding of what was going on and where the mine was at in the recovery effort. There were a lot of operational impacts to the concentrator, smelter, refinery, and tailings as a result of the Manefay failure, and recovery ranged from reduced production rates to highly variable grades that affected downstream operations. By having regular communications, everyone had a better idea of progress as well as the assurance that the recovery work was going forward.

First Ore Team

The initial Cornerstone production was a critical first step in the return to full sustainable production that would make the Bingham Canyon Mine a long-term entity once again. The next step was to start producing ore from the bottom of the pit.

Although a lot of equipment had been damaged or destroyed, there was adequate loading, hauling, and drilling capacity to at least start production. The crusher, conveyor, and tunnel leading to the mill were intact and operational. Therefore, the basic requirements were available to begin ore production, even though it was not enough to reach and maintain the maximum production capabilities of the downstream facilities. However, to start ore production, several hurdles had to be crossed before production could commence. The first hurdle was that a large mass of material called the head scarp still rested on the Manefay bed far above the ore, and it was unknown how stable the material was or if it would affect the ore operations should it fail. Considering the two slides on the night of the Manefay, no one was willing to take a chance of putting people in an area that could be covered up if the head scarp were to fail unexpectedly. Significant geotechnical studies on the stability of the head scarp and potential run-out areas if it were to fail were required to understand the risks before mining could commence. Figure 3.2 shows the location of the head scarp at the top of the Manefay failure. Coincidentally, the photo shows the head scarp being covered by the shadow of a passing cloud, which is fairly symbolic of the shadow that the head scarp cast on the remediation work for several months until it could be stabilized.

Figure 3.2 • Head Scarp

Figure 3.3 • Code 30 Location

The second hurdle was to obtain another modification to the MSHA 103(k) Order. This modification would be much more difficult to get because, unlike the Cornerstone area, the ore was in the bottom of the pit and much closer to the failure area. A detailed plan that included the geotechnical studies and risk assessments was required to get MSHA to amend the 103(k) Order.

Thirdly, an operational plan had to be developed. Only one shovel remained in the pit, and it was not operational because of the preventive maintenance that was being done on it when the Manefay evacuation took place. A plan was needed for making the equipment ready to operate and determining where it would operate as well as who the operators would be to run the equipment.

The final issue was getting access to the bottom of the pit. Most of the Keystone access road was intact, except for a few hundred feet at the bottom of the ramp that had been covered by the Manefay debris. In addition, there was a section of the haul road at the bottom of the pit, called Code 30, that had been covered by the debris. It would have to be cleared to access the areas of the ore that had not been concealed with rubble. Figure 3.3 shows the location of Code 30 in the bottom of the pit in relationship to the 68 shovel (P&H 2800 shovel, assigned no. 68 by the company).

To reach the goal of returning to ore production would take significant effort from many people with a wide variety of expertise. The geotechnical analysis fell on the Geotechnical Team with the support of both contractors and the Rio Tinto T&I experts. The operation plan fell to a multidisciplinary team that was created to look at all operational and safety aspects of returning to production. This group was managed by Tim Juvera, the mine's manager of asset management.

Tim brought together a team that consisted of operations, maintenance, safety, environmental, process improvement, production support, and engineering personnel. Not only did the group cover many disciplines, a wide range of levels within the team were also represented, such as managers, superintendents, supervisors, operators, mechanics, blasters, planners, engineers, and coordinators.

Building a multidisciplinary group not only provided expertise and understanding to solve the issues of starting the First Ore production, it was also symbolic of how the senior leadership wanted the company to work going forward. Tim managed the team so that all members could provide input and have an impact on the plans. For most of the team, but especially the hourly employees, this was a very empowering process. They were working with people from a variety of disciplines and levels within the company, and what they said mattered. No one was allowed to dominate, so everyone contributed. This became the model of not only the First Ore Team, but also how the Operations and Maintenance Teams would work when ore operations resumed.

Key to the work of the First Ore Team was to build plans of what needed to be done, followed by a risk assessment of what could go wrong and how to mitigate those risks. After the risk assessments were complete, the team put together a sign-off document that required signatures from a number of managers and experts before work could be started. This sign-off document was put together by the operations manager at the concentrator. He had also worked as the manager for the Long-Term Planning group at the mine, so he was familiar with the mine's processes.

One of the resources that was brought in to help the First Ore Team (as well as future remediation teams) was Marcel Perreault. Marcel had worked with Stephane Leblanc at a different company and had a strong background in process improvement. Part of Marcel's job was to challenge the goals that the team set for themselves and future operations. Initial estimates of production for the 68 shovel were 50,000 tons per day based on having only one shovel as well as the logistical difficulties of working with the Keystone access road. Marcel asked the question: "Why not produce 75,000 or 100,000 or even 120,000 tons per day?" Many times, the answer came back that higher rates or lower cost could be achieved.

Marcel also improved the sign-off process. When the team first started using the sign-off document, they were frustrated because the people who needed to approve did not get around to reading and signing the document in a timely manner, which delayed the process. I was one of those people who was slow at times in reviewing and signing the plans that had been put together because there were so many other priorities at the time. When the First Ore Team members started to complain that the sign-offs were not being reviewed, Marcel challenged the team by asking why they let people do that. Why was the team member not contacting the approvers and telling them why the approval was important and setting a date (or time) by which the document had to be approved?

The First Ore Team started contacting the slow approvers (including myself), and behaviors quickly changed. No one wanted to be a bottleneck, so everyone started to make it a priority to review the plans and respond accordingly. But this did not mean that people just signed the approval documents blindly. There were several instances where a problem was identified in the plans, which resulted in the plan being sent back to the First Ore Team. At that point, the issuer was the bottleneck, and very quickly the issues would be addressed by the First Ore Team so the approvals could go forward.

In the two weeks between starting production at Cornerstone and the First Ore production, the First Ore Team was putting together plans and risk assessments and the Geotechnical Team was developing the analysis to understand the risk to the First Ore production from a failure on the head scarp. The geotechnical consultant was building models in Clara, a three-dimensional (3D) software program, to understand the stability of the head scarp and in

Dan-W 2D and 3D to understand how far the head scarp would run out if it were to fail catastrophically—like the Manefay had failed. In addition to these programs, rock-fall analysis was being performed on the Keystone access road to determine the risk of opening the access ramp.

On April 17, the Geotechnical Team completed the rock-fall analysis for the Keystone access road and the First Ore Team completed the risk analysis and sign-off document for the dozer work to open the ramp. Nate Foster, one of the operations superintendents, was assigned to walk the documents through the approval process. After submitting the plan, Nate called on all of the approvers, and by the end of the day he had all of the signatures required to go forward. Next, the plan was submitted to MSHA to get temporary approval to use a dozer on the ramp so access could be gained to the 10% haul road below the Manefay failure. MSHA gave their approval and soon the first piece of equipment was working on the first step toward getting to the ore at the bottom of the pit.

By April 21, the Keystone access road was opened and a trip was arranged for the MSHA inspector and various company representatives to take a trip down to the hairpin turn near the Code 30 area to determine what was going to be needed to reopen access into the ore operations. Besides safety, operations, and environmental representatives, I joined the trip to offer an engineering perspective. This was the first excursion to the bottom of the pit by MSHA and most of the managers since the night of the Manefay failure.

The first portion of the trip was interesting because of all the switchbacks going down the Keystone access road. More than half of the 13 hairpin turns required three-point turns (see Figure 3.4) for the vehicle to traverse the route down to where it connected to the 10% haul road. Travelling to the 10% haul road was an uneventful, yet odd, journey. It was not until we got partway down the 10% haul road that I began to realize that the strange feeling was because there was no other traffic on the road. In the mine, there is *always* traffic on the major haul road, but that day was eerily different.

Figure 3.4 • Keystone Access Road

When we rounded the final corner at Code 30, we were able to catch glimpses of where the Manefay had failed, but we did not get the full effect until we exited the vehicle. Once out, it was almost scary. From our vantage point, we could see the debris material that filled the bottom of the pit. Looking up toward the top of the head scarp—nearly ½ mile up and 1½ miles away—the view was absolutely breathtaking! The Manefay left an enormous void where just a few days ago had been solid rock. The dramatic colors of the debris flow going down the slope made a sharp and unforgettable contrast to the normal tint of the pit. Standing there, the hair on the back of my neck started to rise, one of those natural voices that warns us of dangerous situations and to be careful. Looking back and talking with others, it was not just the visual aspect of the Manefay that was spooky; it was also the sound—or lack of it. Working in a large mine, there is always noise. There is always equipment operating somewhere or radios blaring. This time there was almost total quiet, with the exception of our own conversations, and at first no one was saying much. Between the visual aspect of the gigantic landslide rising more than half a mile above and the lack of sound where there has always been noise before, it made for an unforgettable experience.

Figure 3.5 is the view from my first trip to the bottom of the pit before we started ore operations. This location is—surprisingly—600 feet *above* what should have been the bottom of the pit. The Manefay debris is piled up in front, near the bottom of the photo. Although I had seen aerial photos and had stood at the top and looked down from various angles around the pit, this view looking up at the Manefay really had an impact. Walking down that road to the debris pile, I remember thinking, "Oh, my God! What are we going to do?" Bingham Canyon Mine always makes me feel small, but the sight of looking up at the Manefay really brought home just how large the Manefay event really was as well as the magnitude of the recovery efforts to come.

Figure 3.5 • First View from Bottom of the Slide

During the tour, the MSHA inspector decided that work could commence on removing the debris that was covering the corner of Code 30 with remote-controlled dozers. The run-out analysis had not yet been completed by the Geotechnical group, so manned equipment would not be allowed until the analysis demonstrated there was no risk that the miners would be in danger should the head scarp fail.

By this time, the First Ore Team had completed a comprehensive risk assessment, and the sign-off, based on the design from the Technical Services Team, had been finalized so work could begin on the Code 30 hairpin turn. In addition, a special sign-off was developed for the start of remote-control operations as an additional precaution, considering the unusual nature of the work.

By April 21, work had commenced on the remote-control work on Code 30. Progress was being made on the geotechnical run-out analysis, which was completed on April 22. Before manned equipment was allowed in the Code 30 area, another sign-off and MSHA approval were required. Additional documents were added to the sign-off, which included a training program for operators working in the area as well as a response plan in case there was movement on the head scarp and the pit had to be evacuated. By April 23, MSHA gave temporary approval for manned equipment to operate near the Code 30 hairpin area.

Figure 3.6 • No-Go-Zone Berm

After the manned equipment started working on the Code 30 area, they were able to complete their tasks quickly. By April 26, a plan was submitted to MSHA for a permanent change to the 103(k) Order so that ore mining in the E4 South sector (the portion of the ore that had not been covered up) could commence. As part of the plan, a berm was constructed outside the runout area of the head scarp to ensure that no one could inadvertently enter that area. Above the berm, people were allowed to work as they normally would. Below the berm was the No Go Zone where people were not allowed to enter without approval from MSHA and Kennecott management. Figure 3.6 shows the location of the berm where it crosses the Manefay debris field.

The geotechnical monitoring was in place, a new response plan had been instituted, and training of employees had commenced. Consequently, MSHA modified the 103(k) Order on April 26. In the meantime, the Maintenance Team started to work on the 68 shovel by accessing it from outside the No Go Zone on a small, steep road that connected the bench that the shovel was on to the 10% haul road. On the day of the Manefay, the shovel had been backed out of the working face and preventive maintenance work had been started on the shovel. This work had to be completed and the shovel put back together before ore production could resume.

At 10:06 a.m. on April 27, the First Ore Team hauled the first load of ore to the crusher since the Manefay failure had stopped production just 17 days earlier. Figure 3.7 shows that first load being dumped into the in-pit crusher.

Figure 3.7 • First Load of Ore

Once again, ore production was flowing from the mighty Bingham Canyon Mine, feeding the hungry concentrator and smelter downstream. Just as people were surprised when waste production had started in the Cornerstone area less than three days after the Manefay failed, they were shocked that ore production had started so quickly. This was the second major milestone for the Bingham Canyon mine in 2½ weeks—but it was just the start for the First Ore Team.

Exceeding Production Expectations

Once the First Ore Team achieved the first load of ore, their challenges and work were just beginning. Nate Foster took over leadership and was responsible for mining as much ore as possible—while doing it safely. The challenges were difficult. The only entrance to the bottom of the mine was via the Keystone access road. This road was completed just days before the Manefay failure and was so narrow that it only allowed for one-way traffic of light vehicles. There were 13 switchbacks on the Keystone, and even pickup trucks had to make three-point turns to get around the sharp corners.

The First Ore Team would have to get people, parts, and supplies up and down this road if they were to continue mining the ore that they had started to mine on April 27. In addition to the limited access for small vehicles, there was no access for big equipment, such as shovels, haul trucks, or large production drills. At least for the time being, the team had to manage with the undamaged equipment in the pit. Scheduling and timing of getting supplies and people into the pit would be critical to prevent excessive downtime for the few pieces of equipment that were available.

The First Ore Team would have to perform all maintenance and repairs in the bottom of the pit because the equipment could not be transported to the Copperfield Shop in the 6190 Complex. In addition, the Manefay debris had covered the maintenance, fueling, and parts supply areas that had been constructed at the top of the Moly Dome, so this maintenance work would have to be done using portable cranes on a gravel pad, which was feasible but not ideal.

The First Ore Team had one P&H 2800 shovel and a Komatsu WA1200 front-end loader to mine the ore. Under normal circumstances, the average daily production planned would be approximately 60,000 tons per day for the shovel, and the front-end loader would be used in a support role, so no ore production would be planned for that piece of equipment. Eighteen of the 320-ton Komatsu haul trucks were undamaged, so there was plenty of haul-truck capacity for the shovel and loader. With the potential delays from accessibility issues via the Keystone access road and limited maintenance capabilities, the initial plan for the daily production rate was set at 50,000 tons per day by the Technical Services Department for the start of ore production. There were many unknowns at that point, but the rate was at least achievable and served as a starting point for the downstream operations.

Before ore production could start, Marcel Perreault (with direction from Stephane Leblanc) challenged the First Ore Team to increase the targeted daily ore production rate. Both Stephane and Marcel understood that if the rate were set too low, then the team would be satisfied to meet that rate and may not strive to reach higher rates. The question that Marcel started asking was, "What is theoretically possible?"—not "What have we achieved before?" This was a great question because the production from the Kennecott shovels was often limited by the number of trucks that were available or how much production was needed from a particular shovel—both of which lowered the average historic production rates. With one shovel and one loader, these constraints were no longer relevant.

As is often the case, the production from a piece of equipment or part of a team is more a function of expectations or needs of the organization and not what had historically been done. Marcel convinced the team to target at least 80,000 tons per day for the P&H 2800 and to not stop there. The team must try to produce as much ore as possible to maximize the copper production and minimize the impact from the Manefay failure. Some of us thought the target was not achievable considering all the challenges after the slide, but this type of thinking discounts how people are able to solve problems when confronted with a difficult situation.

The First Ore Team did not know how they would reach the higher expectation. Nevertheless, they bought into Marcel's arguments and started making plans to maximize the ore production. The mine was completely different after the Manefay, and Nate Foster, Tim Juvera, and Matt Lengerich knew that the team working on the ore production would have do things differently because of those changes. Before the Manefay failure, the company had a fairly normal structure for the people operating the equipment to report up through the operations manager and the people who repaired the equipment to report up through an entirely different set of leadership to the asset management manager. After the Manefay, the decision was made to change the reporting structure so that all of the operations and maintenance personnel working in the bottom of the pit would report up through one set of leaders and ultimately to Nate.

Because of the Manefay, the maintenance and operations personnel shared common problems, such as how to effectively transport people, equipment, and supplies in and out of the pit. Thus, a bond was created among all personnel working in the bottom of the pit. By having one set of leaders, it was easier for them to strive to meet common goals. One of those goals was to produce as much ore as possible, which meant keeping the P&H 2800 shovel working as much as possible. A year after the event, the First Ore Team still talked about the teamwork and camaraderie they felt when working with the integrated team of maintenance and operations personnel. They credit that opportunity to much of the success they achieved.

In the first four days after starting ore production, the P&H shovel averaged nearly 59,000 tons per day—clearly better than the Technical Services Department had expected, but not yet up to the goal of 80,000 tons per day. However, the Komatsu WA1200 loader averaged nearly 29,000 tons per day, which had not even been taken into consideration in the projection. Consequently, the First Ore Team averaged just under 88,000 tons per day in the first four days of production, exceeding a difficult goal they had set by 10%. And the First Ore Team was just getting started.

Between May and early July, the First Ore Team continued to increase its production rates. In May, the ore production exceeded 94,000 tons per day and a new issue was brought to light. The original plans showed that there was enough ore available to mine until nearly the middle of 2014. However, the First Ore Team was producing ore nearly twice as fast as originally predicted, and that rate was continuing to increase. Based on those rates, the mine would run out of ore by the end of 2013 or sooner, thus the ore could run out well before more could be uncovered by the remediation work. Something had to be done, otherwise all of the great work that the First Ore Team was doing to increase production would be for naught because there would be a gap in production by the end of the year.

Once again, a challenge was put before the teams: find a way to prevent a gap in ore production. The answers to the problem were a joint effort. The Mine Planning Team determined that additional ore could be uncovered sooner if some of the debris in the bottom of the pit could be mined. To solve that problem, a second shovel was being constructed in the bottom of the pit to help increase production. Ultimately, the First Ore Team would have to produce even more every day.

In the middle of June, the mine plan was modified to start mining the debris material in addition to the ore. More than 4 million tons of waste would have to be mined to ensure that the mine would not have an ore gap. During the month of June, the ore production had increased to nearly 112,000 tons per day, and good progress was being made on the construction of a new Hitachi 5600 hydraulic shovel (no. 64 shovel) in the bottom of the pit. The P&H 2800 was averaging nearly 100,000 tons per day—double the original assumptions.

In early July, the new shovel was assembled and the average production increased to 142,000 tons per day of waste and ore, with the two shovels and the front-end loader in operation. Finally, just before October of 2013, the mine was able to move a second hydraulic shovel into the mine while they were constructing the 10% haul road, and the production rate jumped again, to 164,000 total tons per day in October. By this time, the mine was close to maximizing the throughput of the crushing and conveying system because of the amazing job the team did in maximizing shovel production. Figure 3.8 shows how the production increased in the bottom of the pit over time.

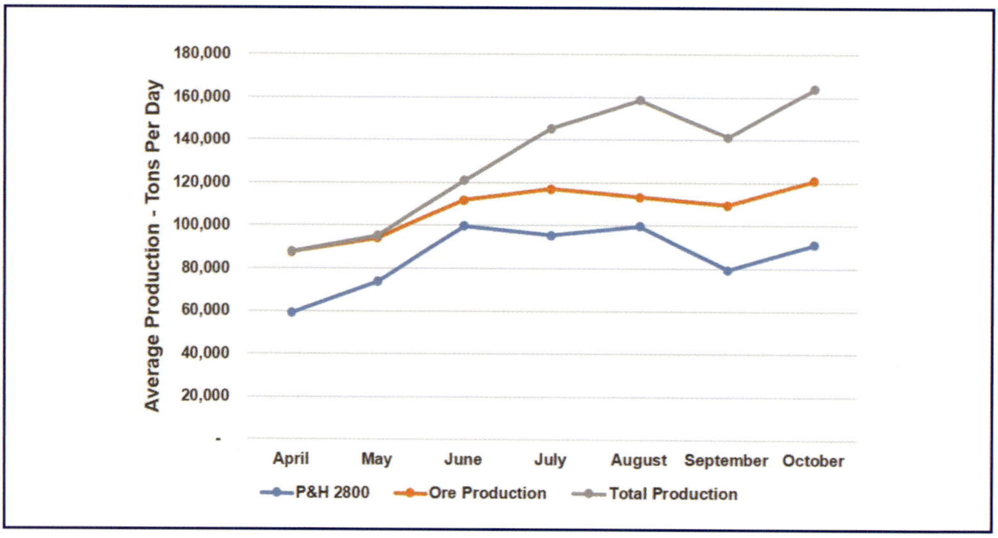

Figure 3.8 • Pit Bottom Average Daily Production

The First Ore Team did a terrific job of using the equipment, resources, and people they had available. They definitely rose to the occasion and demonstrated the shovel production that was truly possible, even in difficult conditions. Figure 3.9 shows the 68 shovel loading a haul truck after the Manefay.

Besides the high expectations and strong teamwork (or perhaps because of these), the First Ore Team changed a number of tactics to achieve these remarkable records. One of the changes they made was to have the mining team work longer hours each day. As additional time was required to move people in and out of the pit, crews worked longer shifts to minimize the time equipment was idle during crew changes.

Figure 3.9 • 68 Shovel Loading a Haul Truck

The time to get up and down the Keystone access road was also a limiting factor, so the First Ore Team scheduled the movement on the Keystone access. The crews coming on and off their shift had the highest priority, so they would not have to wait for traffic on the road. This also minimized time for changing out the crews.

Given that three shovels had been destroyed, the First Ore Team understood that operational loading equipment was their greatest bottleneck to increasing ore production. Consequently, they maximized every minute they could to keep the shovels operating. To minimize delays, they assigned two operators for the P&H 2800 shovel. If one shovel operator was tired or needed a break, the second operator was there to take his or her place with minimal interruption in production.

Furthermore, the First Ore Team treated any breakdown of two hours or longer of the 68 shovel as a major incident. If the breakdown lasted longer than two hours, an incident commander was assigned to get the shovel operational as quickly as possible. This included almost unlimited spending authority. Considering that all of the downstream plants were relying on the production from this one piece of equipment—especially for the first three months after the Manefay—every time the shovel went down, it *was* a major incident. This policy started within a month of the Manefay when the shovel lost a major gear. When the Mine Management Team reviewed the time it took to get the part repaired using the normal approval process, they decided that the procedures needed to be changed. By using the techniques that were in place to manage major incidents, the approvals could happen much faster. This fast-track process was not used often, but when it was implemented, the result was very effective. Figure 3.10 shows the intermediate and main hoist gearing that was damaged on the 68 shovel, which led to the First Ore Team to change their procedure so that major breakdowns were treated as a major incident.

Figure 3.10 • Gear for 68 Shovel

Developing a New Mine Plan

Mine plans are the road map for a mining operation. A mine plan includes the design of the mine, which comprises the physical dimensions of highwalls, benches, and ramps, as well as the location of ore versus waste. It dictates what equipment will be used and the expected production from the equipment. The mine plan also includes a schedule of what material will be mined from each location in the mine by which piece of mining equipment for each time period. It optimizes the mining process to minimize risks and maximize profit, from the very short term through the entire life of mine. Without a mine plan, the operations would be working blind and would be at risk of optimizing certain aspects of the mine to the detriment of the overall operation.

The Manefay failure was so large that in less than two hours it had transformed the face and surface of the Bingham Canyon Mine. This transformation made the previous mine plans obsolete and useless. A completely new plan was needed for the mine to go forward, and this mine plan was needed as soon as possible to ensure that operations were doing the right work. Even the most basic questions had to be answered with a new mine plan, such as, is the continued operation of the Bingham Canyon Mine still economical?

Developing this new mine plan would not be fast or easy. Most of the previous mine plans at the Bingham Canyon Mine were incremental in nature since they were modifications of the previous plan. The changes from the Manefay were so massive that planning had to completely start over. The surface was dramatically changed and the Manefay exposed new risks of highwall failures, so new designs would have to start from scratch. Three shovels and several haul trucks were destroyed, so the equipment selection had to be reconsidered. Even the equipment productivities were fluctuating, given that the First Ore Team was achieving productivity with shovels that had never been achieved before, which put the scheduling process in flux.

The mine planning effort started in the weeks before the Manefay when Cody Sutherlin, the superintendent of mine planning, started training some of his mine planners on pit design work. Cody had moved to the mine only a few months earlier and even though he was the superintendent, he was doing a majority of the design work for the short- and medium-term mine plans. When Cody realized that the Manefay would likely create the need for new plans and he would not be able to perform all of the design work himself as well as manage the rest of the planning effort, he started to guide other engineers to do design work. In the few weeks before the slide, Cody developed Jon Heiner and Josh Davis to become proficient in mine design work. By the time the Manefay actually failed, both were capable design engineers, which was critical for the massive design effort after the failure.

On the day of the Manefay, Cody organized and managed the Mine Planning Team's move to the RTRC in Daybreak, where the team was scattered over two wings on the second floor of the building. The Production Support Team, who needed to work closely with the mine planners, were located in a different part of the building. Although not consolidated, everyone at least had a place to work and they were able to get their computers set up and operational. It was impressive that the entire group could pick up from one office at the mine on Wednesday and be set up and operational in a completely different office the next day. When evacuating the mine, computers were carried out to avoid exposure to a potential dust cloud. It was fortunate that the computers were removed, not because of the anticipated dust, but because the computers with planning programs were needed to begin the planning efforts immediately.

On the morning of April 11, Jon Warner saw the pictures of the Manefay failure and knew the Mine Planning Team would need help. Jon was a mining engineer that was in the Mine Planning Department a few months earlier but had been transferred to Operations as a shift supervisor. Since Jon was a supervisor in the bottom of the pit, he figured he would not be needed in that position for a while, so he went to the RTRC to offer his help to the Mine Planning Team. Cody knew that he would need additional help and immediately accepted Jon's offer. Jon not only had mine planning experience, he also had strong communication skills that would turn out to be extremely valuable for the team going forward.

Cody was filling in as the manager of Technical Services Mine Planning on April 11th and most of the 12th because I was either in Incident Command or attending the Mine Management meetings. During that time, the team was estimating the size of the failure. Jon Heiner did a quick calculation of the size by the morning of April 12 and estimated the failure to be 141 million short tons. Cody communicated that the failure was estimated to be between 140 and 150 million short tons. I in turn used an estimate of 150 million short tons when communicating to the rest of the Mine Management Team. Estimates sometimes take on a life of their own in crisis situations, and someone assumed that the 150 million tons was metric tons, so the estimate became 165 million short tons. This was reported to the press, and from that point on, most of communications pertaining to the size of the Manefay were given as 165 million short tons.

By April 18, aerial photographs of the Bingham Canyon Mine had been taken and digitized, so an accurate estimate of the size of the Manefay could be made. The volume of the failure was calculated with two methods, and the difference between the two was less than a million cubic meters. They averaged 66.2 million cubic yards. Based on the typical density of rock at the Bingham Canyon Mine, the official tonnage of the Manefay was then calculated to be 144.4 million short tons—extremely close to the initial estimate.

After the Manefay failure, Eric Eatherton took over as the official manager of the Production Support Team so that I could focus on managing the post-Manefay planning effort. Several members of the Production Support Team were key to the planning effort and worked as an integrated part of the Manefay Planning Team. Formal reporting relationships were not as important as getting the job done, and the groups worked seamlessly together.

The Production Support Team was also responsible for many of the logistics after the Manefay. This included tasks such as acquiring lease vehicles for the remediation supervisors and finding work space for staff at the mine offices. One of the first requests of this group was to find a large work area in the RTRC so that the teams working on the Manefay planning could work together—as recommended by Anna Wiley. Initially they found 25 cubicles and two offices by working with the regional site manager, Kathy Box. After more discussion it was determined that at least 50 cubicles would be needed. Amazingly, by the next day, arrangements had been made for the team to have the majority of a wing with 50 cubicles and three offices in one area. For this to happen, many people that had offices on that floor had to move to a different area—a real disruption to their work. By April 17, the entire team working on the Manefay planning was moved to the third floor in the RTRC. The layout was extremely effective and the ability for the team to work together was one of the factors that led to the ultimate success of the planning effort. I will always be appreciative to the Production Support Team, site management, and all of the people that moved offices so that the Manefay Planning Team could work together. It was that type of willingness to work together and sacrifice that made the recovery effective and successful.

On the afternoon of Friday, April 12, there was a meeting of the various groups that would be doing the post-Manefay planning, including Technical Services Mine Planning, Production Support, as well as Rio Tinto Technology and Innovation (T&I) to discuss the work that was needed going forward. Mark Button of the Rio Tinto T&I group attended the meeting since his team's help would be needed in the planning effort. During the meeting, more than 70 tasks were generated as well as lists of resources and tools that would be needed to complete the work. The workload was large and covered everything from locating engineers that could be utilized from other sites within Rio Tinto to the various plans required as part of the remediation process. Although the work was great, time was short and resources were limited—at least at that point. But the team was motivated and had grown in capability and confidence during the weeks before the Manefay. They had already proved they could get an amazing amount of work done in a short period of time.

Many of the people on the Mine Planning and Production Support Teams had been putting in a great number of hours over the previous few weeks. After the Manefay, these work requirements were not going to diminish and everyone had to work the first weekend following the failure. It was one of the very few times that I had asked everyone to work a weekend. As it turned out, a vast majority would have worked that weekend anyway—as they did many other weekends for much of 2013—not because they had to, but because they understood the importance of the work they were doing and the dedication required to get the job done.

Cody Sutherlin took on the responsibility for supervising the short- and medium-term planning work. Joan Danninger from Rio Tinto Strategic Production Planning group was hired by the mine to supervise the long-term planning work. Joan started the week after the Manefay, and she had no chance to catch her breath because the mine required a complete set of long-term plans as well as an evaluation of the ultimate pit shell by the first week in July, less than 11 weeks away.

By April 19, planning was moving forward at a fast pace. Plans had to be made for multiple time frames, including short term (weekly to one month), medium term (monthly for two years, then yearly for three years), and long term (yearly for life of mine), as well as a new pit shell evaluation.

As with all parts of the Manefay, communications were essential to success. Because the Manefay was so massive, it was very difficult to grasp what was needed when looking at photos or maps. It was hard to prioritize or even to logically explain to people all the work that was necessary. It was overwhelming to look at a photo showing all the destruction that the Manefay had done to the mine and see a path to recovery.

The answer to communicating the work that needed to be accomplished came from Jon Warner, the engineer who offered help just after the failure. In addition to Jon's skills as an engineer, he is also an artist with a good understanding of colors and what people see when they look at a drawing. Jon took a computer figure of the mine after the failure and color coded the different parts of the mine that would need to be remediated. Figure 3.11 shows one of the first versions of the drawing that Jon had put together to show all of the steps required to remediate the Manefay. Anyone looking at the drawing could almost instantly understand the work that needed to be done. Instead of one large area of destruction, Jon's drawing enabled a person to see where the scarps needed to be remediated and where safety benches had to be cleared, as well as how long it would take to do the work. This simple drawing changed how people saw the Manefay, and suddenly it could be broken down in manageable pieces that could be addressed instead of a large mass that was overwhelming. Without the drawing, it would have been nearly impossible to adequately describe what people needed to do to return to full operation.

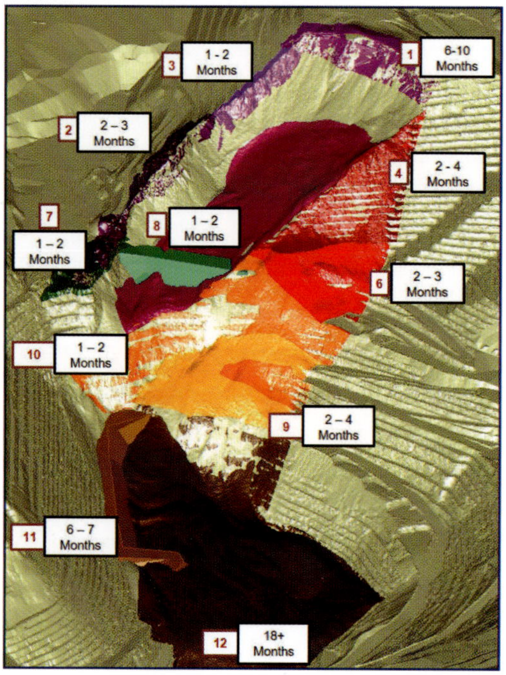

Figure 3.11 • Manefay Remediation Steps

One of the first orders of business was to find additional resources to help with the planning effort. After the earlier planning meeting on April 12th, Cody and Joan had reached out to the Rio Tinto T&I group, to engineers that had worked at Kennecott but moved to other locations, and to several consulting firms. The initial plan was for the T&I group to take on the planning for the scarp remediation and rebuilding the 10% haul road as a separate project, but the T&I members were soon integrated into Cody's team. No less than 13 engineers, geologists, and geotechnical engineers supported the remediation planning work from the Rio Tinto T&I group. Their help and support was invaluable after the Manefay and shows the benefit of being a part of a large company like Rio Tinto that has internal experts.

Two engineers from other Rio Tinto mines were sent to help on a temporary basis: Brandi Nobis from the Iron Ore Company of Canada and Dario Bernabe from Oyu Tolgoi in Mongolia. Mining engineers were brought in from three engineering firms as well as three software vendors to help with engineering support and training during the initial mine planning process. Nearly 50 people were involved in the mine planning process after the Manefay.

There were multiple priorities and projects for the planning group in the weeks after the Manefay that included the following:

- Mine designs to support the First Ore Team, remediation plans, risk assessments, and MSHA 103(k) modifications that were primarily done by Jon Heiner, Josh Davis, Jon Warner, and Ryan Betts (from the Underground Team) and managed by Cody Sutherlin. The number and speed of the designs were impressive.

- Short-Term Planning to support the day-to-day ore and waste production that was performed by Geoff Bedell (senior blast design engineer), T.J. Gillespie (ore controller), Eric Hoffman and Braden York (graduate engineers), and managed by Cody Sutherlin. The engineers were located in buildings outside the mine instead of the RTRC and had to work independently with little oversight. This was a difficult job because the operations were performing at record-setting levels and that meant the plans were continually modified to match the new rates.

- Mine designs to support the medium- and long-term mine plans. This was a comprehensive design effort that not only involved the engineers doing the remediation plan, but it also included Chris West, Bill Rose (a consultant with a long history at Kennecott), Dario Bernabe, and Ed Woods (a new engineer that transferred from Australia). Cody Sutherlin managed this effort. They made designs for multiple options and pit configurations.

- Medium-term schedule for the budget of which the initial results were due in early July. This work was primarily performed by Denee Hayes, Ryan Mauser, Brandi Nobis, and Nate Armstrong. Cody Sutherlin managed the team. This was the first time that the budget would be done using a new scheduling program, which added considerable challenges and work, and support from the vendor.

- Long-term schedule for the life-of-mine plan to determine the value of the Bingham Canyon Mine. Mike Watson was the key person working with the schedule, with support from Ricardo Garcia and John Finch (from T&I), Chris West, Nate Armstrong, Karl Mudge, and Bill Rose. The long-term work was dependent on both the short- and medium-term work and was managed by Joan Danninger.

- Pit shell evaluation to determine if the previous ultimate pit was still applicable after the changes from the Manefay, performed by Mike Watson and supported by an outside consultant who was proficient with a program that utilized the Lerchs–Grossman algorithm to optimize the ultimate size of the mine. Joan Danninger managed the work.

With the Bingham Canyon Mine being one of the largest and most complex open pit mines in the world, the mine planning effort requires world-class planning software and tools to be able to perform quality designs, schedules, and optimization. The best planning tools were required to manage the sheer magnitude of data generated by the size of the operation, the number of drill holes, samples, technical analyses, and pieces of equipment that are all a part of the mine plan.

The planning effort was started with a program called MineSight by Hexagon Mining, which is used to input, store, organize, and manage the incredible amount of data available from the Bingham Canyon Mine. These data include geologic, geotechnical, metallurgical, as well as operational information in databases. These data can then be used with programs internal to MineSight or by external programs that are built by other vendors. Having a single platform to store all the various types of data is invaluable to the planning effort for a mine such as Bingham Canyon.

In addition to storing the data, MineSight is an integrated program used to develop the mine design of where and what material will be mined. The program has built-in 3D computer-aided design tools to allow the engineers to virtually build the mine in advance of physically moving the rock. The MineSight program is used for short- and medium-term designs and analyses at Bingham Canyon Mine.

Before the Manefay failure, the Mine Planning Team was in the process of implementing a new model called MineSight Schedule Optimizer (MSSO) to turn mine designs into schedules that included the amount of material to mine by time period, equipment hours to mine the material, as well as the resulting qualities of the ore to be mined during the period. Unfortunately, the implementation was not completed beforehand. Even though mine planning was critical after the failure, the decision was made to complete the implementation, which was somewhat risky considering the critical nature of the timing. To help with the implementation, MineSight supplied an application engineer, Abinash Moharana, to support the implementation effort. Implementation of new software can be tedious at best, but the team, led by Denee Hayes with Abinash's help, was able to complete the implementation for the July time requirement. MSSO was used to schedule the first two years of the budget plan. MSSO's value was with the month-to-month detail provided by the program that is useful for operations.

One of the first tasks for the mine planners after the slide was to determine if the ultimate size of the mine should change when compared to pre-Manefay. To perform that task, the team used a program called Whittle by Geovia. Whittle uses the Lerchs–Grossman algorithm to determine the economic limits of a mine, and it uses the MineSight database to perform this analysis. The result was that even though the Manefay had significantly changed the face of the mine, the ultimate mining limits remained unchanged.

A third software program, Comet by Comet Strategy, was used by the team to predict the budget from year three onward as well as the life-of-mine plan schedules. Comet's value was that instead of using an in-depth analysis of data to create a monthly schedule, it could quickly calculate an annual schedule. Results can then be optimized by performing multiple iterations while running overnight in a batch mode. This speed was essential for the life-of-mine work because of the numerous scenarios and amount of optimization work that were required to be performed since the mine plan work was starting from scratch. Comet not only calculated the schedules for the life-of-mine plans, but it also calculated mining costs and estimated net present value for the mine, which was critical when making comparisons between various scenarios. Just after the Manefay, the only person on the team that had background in using Comet was Joan Danninger. Brett King, the founder of Comet Strategy, was brought in to provide additional training, and by the time the first life-of-mine plan was completed, other engineers on the team were capable of using the program. In particular, Mike Watson became one of the top users of the program. Comet also used the MineSight databases and was invaluable to the success of the team by accomplishing hundreds of analyses in a relatively short period of time.

The pace of engineering work was incredibly fast after the Manefay failure. New designs for the remediation work and mining of the ore were updated almost daily. By the end of April, the long-term planners had determined that the optimum ultimate pit for the Bingham Canyon Mine was unchanged from before the Manefay, which was good news, given that the mine designers had gone forward with that assumption and were making progress on the life-of-mine designs.

On May 6, the Manefay Planning Team had to make a major adjustment to the group's management. Cody Sutherlin was promoted to manage the Next Ore Team, which was responsible for the remediation work on the Manefay. This was a great opportunity for Cody, but it put the Planning Team in a bit of a bind. Cody had been managing a large part of the work because he had both the technical and leadership skills to work on a variety of priorities at once (which is also the reason he was chosen as the Next Ore manager).

Ultimately, Cody's responsibilities were divided up for three people. Geoff Bedell agreed to take on the responsibility for the Short-Term Planning Team. Geoff was always happy to do the blast designs and senior engineering work and had often stated that he never wanted to be a superintendent. But when asked to fill in, he did not hesitate because he understood the need and was willing to step up.

Jon Warner was made an acting superintendent and given the responsibility for all design work for the medium- to long-term plans. The design work for the remediation was moved to the Next Ore Team, so Josh Davis was assigned to that group. Jon's strong communication skills and natural leadership capabilities made him a great choice for the position, even if this was relatively early in his career.

Denee Hayes was chosen as acting superintendent for medium-term scheduling. Denee was the most familiar with this area, which was critical because of all the efforts to institute a new scheduling program.

Joan Danninger continued on as the superintendent for the Long-Term Planning group. Over the next eight weeks, these individuals worked together to complete the designs and schedules to deliver the initial life-of-mine plans by July 10. It was a significant achievement, considering that the mine planning effort started with a nearly blank sheet on April 10 and was the culmination of many long days and more than a few weekends.

The initial budget and life-of-mine plan was just the start for the Manefay Planning Team. Once that work was done, the team had to improve and optimize the plan. More was being learned everyday about what needed to be done to remediate the Manefay as well as understanding the speed and capability of the operations as the Operations Team continued to break productivity records.

From the month before the Manefay until four or five months after the failure, the Mine Planning Team went through a tremendous transformation. Before the Manefay, the medium- and short-term mine designs had to be made by the superintendent because no one else had the required skills. After the Manefay, the technical and communication skills of the team were substantially greater. The team rose to the occasion and was capable, confident, and much more effective because of the incredible amount of work and planning that had to be done. This work serves as a tribute to the tremendous dedication of all members of the team.

New Equipment

Recovering from the Manefay required a wide variety and numerous pieces of new equipment—and it all needed to be operational as soon as possible. This new machinery would not only replace pieces that were destroyed in the Manefay, additional equipment was also needed for remediating the largest geotechnical failure in the history of mining. The amount of required new machinery could have easily fully equipped a good sized mine. Unlike a new mine, though, the new equipment was needed in weeks to a few months instead of the year or more that would normally be required to source this number of items.

The heavy mining equipment with the longest delivery time included 1 electric shovel, 1 hydraulic shovel, 4 blasthole drills, and 20 haul trucks. A huge number of smaller equipment was also required, primarily for the remediation work, and included 38 dozers, 15 excavators, 3 front-end loaders, plus a variety of rubber tire dozers and graders.

In addition to the heavy mining equipment, the mine required a large number of smaller and support equipment. Much of the early remediation work would be done in hazardous areas, necessitating the purchase and installation of 25 remote-control systems. Smaller replacement equipment ranged from forklifts and cranes that had been destroyed in the bottom of the pit to new bucket trucks and fuel trucks that would be needed to operate the remote-controlled equipment.

Nearly 180 pieces of equipment had to be sourced, ordered, delivered, and commissioned. Some of that equipment, such as the shovels, were long lead-time items that would normally take a year or more to go from sourcing to operation, but this equipment was critical to get the mine back to full production. Other items, such as the dozers and the remote-control systems, were critical path items required to do the remediation work. Any delays in purchasing of the critical path items could result in a gap in ore production later in the year and would have a major financial impact. Clearly, the procurement of equipment was a critical task not only because of the amount of purchases that needed to be made, but also because the equipment was critical for the mine to meet the tight schedules to be successful.

The process of sourcing the equipment started almost immediately. On April 12, Evan Hill prepared an inventory that identified the shovels, loaders, and drills that were available to be purchased at that point in time. Evan was part of Rio Tinto Procurement and manager of service delivery and procurement for the Kennecott operations. This list included a P&H 4100XPC AC electric shovel that had been purchased by another Rio Tinto mine and was in the process of being loaded on a freighter, but could potentially be diverted for delivery to the mine (Figure 3.12). The list also included a Hitachi 5600 hydraulic shovel that was just four weeks from being built and available, as well as two LeTourneau 1350 loaders that were available immediately or in the near future. There was other equipment on the list as well, but they did not meet the time requirements or the specifications that were needed. In less than two days after the Manefay, Rio Tinto Procurement located two shovels and two loaders that were available almost immediately. Quickly finding the longest lead-time equipment was very impressive and ultimately critical to the 2013 ore production.

The sourcing, procurement, and commissioning of the rest of the equipment would take a much larger effort. A procurement regional manager with knowledge, experience, and vendor contacts as well as resources inside Rio Tinto was assigned by the Rio Tinto Procurement group to lead the purchase of the heavy mining equipment.

Figure 3.12 • New P&H 4100XPC AC Shovel

The sourcing and procurement of the heavy mining equipment was only one piece of all the work that needed to be done to supply the mine with the required machinery. There were many smaller pieces of equipment that were necessary to keep the mine and remediation going, plus the logistics of delivery, commissioning, and making sure that the equipment had appropriate supplies, such as fuel, to operate. Many of these tasks fell on the Production Support Team.

Jessica Sutherlin, the senior analyst on the Production Support Team, took on the task of leading the equipment procurement and logistics for the team. Much of the Production Support Team was working in the RTRC, side by side with the Manefay Planning Team. The close working environment was beneficial given that the Manefay Planning Team was determining the equipment requirements and the Production Support Team was responsible for working with Rio Tinto Procurement and organizing the overall acquisition of the equipment. Having the two teams work together in the same work area created an environment with very good communications and understanding of needs and progress between the groups.

Jessica reported to Lori Sudbury, the superintendent of Mine Monitoring and Control, and the two of them with the support of Eric Eatherton, manager of the Production Support Team, put together a strong group to work on procurement and logistics of the smaller equipment and supplies. Many on the team were young, but also energetic and innovative. The team included the following members:

- Joe Stephenson, who came from Rio Tinto Procurement and understood the procurement process
- Chris Haecker, a mechanical engineer, who did terrific work to get the Hitachi 5600 (no. 64 shovel) down the Keystone access road and constructed in the bottom of the pit
- Eric Cannon, who came from the Operational Readiness group and put together equipment specifications and helped to commission equipment
- David Powell, a senior engineer, who worked closely with the Manefay Planning Team to determine equipment needs as well as putting together historic equipment performance data
- David Olson, a reliability engineer, who normally worked in the downstream plants but quickly became a mining equipment expert
- Karen Bakken, who was responsible for finding ways to deliver fuel and supplies to the bottom of the pit
- Brian Coates, a mining engineer, who helped with commissioning equipment
- Sunny Konduru, who made the arrangements so the Mine Planning and Production Support Teams were able to work together at the RTRC

Of course there were many other people involved who were vital for getting the new equipment to the Bingham Canyon mine in a timely manner. None were more important than the vendors themselves. All of the vendors understood the critical nature of getting the equipment to the mine as quickly as possible and went above and beyond to help. Often, the agreements were verbally made and the vendors would start shipping equipment before they received the signed paperwork.

By April 22, the negotiations were completed on the P&H 4100XPC AC electric shovel and Hitachi 5600 hydraulic shovel. The requisitions were approved shortly afterward and the shovels were ordered—in just over twelve days since the Manefay failure! Ordering very expensive shovels in such a short period of time was a significant feat, considering the normal approval time for Rio Tinto.

Recognizing the necessity to get things done quickly, the senior leadership in both Kennecott Utah Copper as well as Rio Tinto's Copper group changed the approval procedure so equipment could be purchased promptly. One of the modifications to the process was that replacement equipment and remediation equipment did not have to go through the standard Kennecott Investment Committee process. The requisition approval limit for Matt Lengerich and senior leaders in Kennecott was temporarily increased to relatively high amounts, given their respective roles. This did not mean that the financial and senior leaders in Rio Tinto were not aware of what was being purchased. Jessica was sending daily reports to the Accounting and Financial Teams. A lot of verbal communications were also going on to keep people informed. It was fortunate that the Mine Planning and Production Support Teams were located on the third floor of the RTRC, as that was the location of the Accounting, Legal, and Senior Leadership Teams as well. The incidental communications that resulted from all of these groups being in close proximity supplemented the official communications and provided greater understanding of what was going on within the teams.

The process changes were important for moving quickly, but there were other cultural changes going on that were facilitated by the Senior Leadership Team. More than once I heard Stephane Leblanc, then the chief operating officer, tell the Mine Planning and Operations Support Teams, "I don't care if you make a mistake. I don't expect you to be perfect. But learn from it and fix it, but do it quickly." The message was clear—speed was more important than perfection.

This thought process was not typical for Rio Tinto capital purchases and reinforced the resolve that people felt for getting the mine back to full production. As it turned out, the sense of urgency that the Senior Leadership Team was creating paid off. There were only a few mistakes made, because people were focused on doing the right thing, and the timely purchase and commissioning of the equipment was critical for the remediation being completed months earlier than anticipated. Delays in the purchase of the equipment could have delayed the remediation and resulted in long gaps in ore production. That result could have cost the company hundreds of millions of dollars.

One of the most obvious successes regarding the equipment procurement work was the acquisition of the Hitachi 5600 shovel. After the Manefay failure, ore production was basically limited to one electric shovel—a P&H 2800 (68 shovel)—and a front-end loader. The 68 shovel was one of the older and smaller shovels in the Bingham Canyon Mine's shovel fleet. The 68 shovel performed amazingly well considering its age and size. But having only one shovel producing ore represented a tremendous risk to the mine. Firstly, if the shovel broke down for an extended period of time, ore production would come to a near halt and downstream operations could be shut down. Secondly, to keep ore production going through the end of the year, some debris material had to be removed in the pit bottom. If the 68 shovel was used to remove the debris material, ore production would stop and total ore production would be reduced.

The potential solution to the 68 shovel risk came as a result of the almost immediate purchase of the Hitachi 5600 hydraulic shovel—but the shovel had to be constructed at the bottom of the pit to be useful. Chris Haecker took on the job of finding a way to get the shovel down the Keystone access road, which was the only access after the slide. Because of its 13 switchbacks and extreme narrowness, building the shovel at the top of the pit and tramming it to the bottom on the access road was not possible—it would not physically fit. The next option was to take the parts of the new shovel down the Keystone access road and build the shovel at the bottom of the pit.

Figure 3.13 • A40 Flatbed Truck

Chris contacted Hitachi to get the dimensions of the largest components to the shovel. From there, Chris worked with Arnold Machinery, the local distributor of Hitachi, to brainstorm various options to get the shovel components to the bottom of the pit. Chris and his team came up with two solutions. The first was to modify a 40-ton articulated Volvo A40 haul truck to a flatbed truck. Chris measured the radii of all of the switchbacks and determined that the A40 could negotiate the corners to reach the bottom of the pit. Figure 3.13 shows the modified truck hauling parts of the shovel down the Keystone access road.

Figure 3.14 • Shovel Frame on Skid

The A40 was able to get many of the parts and pieces down into the pit quickly and efficiently. There were some components (such as the crawler assembly, frame, and housing) that were just too big for them to be moved safely down the access road on a truck. To transport these larger components, Chris's team utilized a method that mines sometimes employ in remote and difficult locations; they built a skid and used two dozers to move the parts down the Keystone access road. Dozers were used because they can pivot on a point when turning and therefore can accommodate the sharp corners. Two dozers were needed so that one could pull the skid down the ramp and around the corners. The skid was connected by a cable to the second dozer as well, which basically acted as a brake so the skid could not slide into the lead dozer going down the ramp. Figure 3.14 shows the two dozers working together to skid part of the Hitachi 5600 frame down the Keystone access road. This part of the road was particularly wide. In other areas, the road was barely wide enough for the dozers to pass.

Not only did the transport team have to be careful in moving the components down the Keystone access, they also had to take the parts in a specific order. The Construction Team had a very limited area to build the shovel (less than half the normal space). They scarcely had room to build the shovel with the components in place and in the order they were needed. Figure 3.15 shows the construction area and the limited space where the team had to work.

By July 2, 2013, just 83 days after the Manefay, the new Hitachi 5600 (no. 64 shovel) was constructed and ready for production. This was an outstanding achievement in itself, but combined with the fact that the shovel components had to be transported down the narrow Keystone access road with its 13 switchbacks and constructed in an area that was half the normal construction site size made the accomplishment even more impressive. This feat was only realized because of the combined efforts, dedication, and innovation of the Procurement group, Production Support group, the manufacturer, and the local vendors. Figure 3.16 shows the newly assembled shovel. Pictured in front of the completed shovel from left to right is Chris Haecker from the Bingham Canyon Mine, Josh Pierce, Warren Christenson, Ron Yocom, Paul Prieto, and Tim Taggart from Arnold Machinery.

Another critical area for returning the mine to full production was the acquisition of the dozers and excavators that were needed to do the remediation work. Delays in finding, delivering, and building that equipment would delay the remediation work and could ultimately create a gap in ore production later in 2013 or 2014.

Almost overnight, the mine suddenly needed 38 large dozers and 15 excavators. No one dealership would have anywhere near that many large dozers and excavators in stock, and time was critical. To complicate matters, most would be leased instead of purchased. Caterpillar and the local dealer, Wheeler Equipment, stepped up to source as many dozers and excavators as quickly as possible. A nationwide search was started almost immediately, and

Figure 3.15 • Shovel Construction Area

equipment was soon being delivered from all parts of the country. Jessica Sutherlin and the team working on procuring and commissioning the equipment tracked the status of each piece of machinery every day to minimize delays.

The first D11 dozer was delivered and operational on May 1st, and the first excavator was ready on May 31st. This was remarkably fast, especially considering that it not only included delivery of the equipment, but all commissioning to meet Kennecott and Rio Tinto standards. Meeting those requirements was not easy because the equipment was shipped from a variety of locations without a typical set of options, so basically each piece of equipment had a different set of safety and production accessories that had to added. No shortcut would be accepted, and all of the fire-suppression systems, two-way radios, dispatch systems, and terrain and remote-control features had

Figure 3.16 • Hitachi 5600 Hydraulic Shovel

to be properly installed and working before the equipment would be released to operations. For the excavators, this meant developing totally new remote-control systems because there were no off-the-shelf equipment for that type of machinery.

By the first week in July, all of the 38 dozers and 15 excavators (not to mention a large number of smaller equipment) had been commissioned and were ready to work. Equipment would not be the bottleneck to remediating the Manefay.

Trucks were another critical piece of equipment that was required to return the mine to full production. Not only would new trucks be required to replace the ones that had been damaged, but the mine now had more than 100 million tons of debris in the bottom of the pit that would have to be hauled out of the pit on the longest haul in the mine.

Figure 3.17 • First New Haul Truck Delivered

It was decided that 20 new 320-ton haul trucks would be required to replace the damaged and destroyed trucks and haul the extra material.

Unlike dozers and excavators, existing inventory of new haul trucks was not available. New trucks would have to be manufactured at the factory. Komatsu was chosen to provide the haul trucks. The amount of time to manufacture and deliver the trucks to the mine was critical.

By June 30th, Komatsu had manufactured, delivered, built, and commissioned the first of 20 haul trucks. This was a very short period, considering they were not selected until early May. As the trucks were being delivered, it was decided to construct some of the trucks in the bottom of the pit. Just after the Manefay failure, the belief was that there was an adequate number of trucks in the bottom of the pit to support the one shovel and loader in the ore operations. However, the First Ore Team was far exceeding the expected ore production, and Chris Haecker and his team were able to construct the Hitachi 5600 hydraulic shovel in the bottom of the pit, which would increase the demand for truck power. To ensure that the haul trucks would not become a bottleneck, four of the new haul trucks were constructed in the pit bottom. The same procedures were implemented that utilized dozers and a skid to transport the components down the Keystone access road. Once the parts were down in the pit, the trucks were constructed on the same pad built for assembling the Hitachi shovel.

On July 26th, the first new Komatsu 930SE was operational in the pit bottom. Over the next three weeks, three more haul trucks were constructed, so the ore operations crew had four new trucks. The ore operations were still limited because of the three shovels that were destroyed, but the First Ore Team was finding ways to increase production. By having the additional trucks, haulage of the ore would never be the bottleneck to production. Figure 3.17 shows the first Komatsu haul truck that was delivered after the Manefay failure.

By October 31, all 20 haul trucks were in operation—less than six months after the contract was signed to purchase the trucks. These haul trucks were critical in Bingham Canyon's ability to maintain high levels of production during the recovery from the effects of the Manefay as well as returning the mine to full production—safely and quickly.

Manufacture of the haul trucks by Komatsu was indicative of the exceptional response by all of the equipment vendors that supplied equipment for Bingham Canyon Mine. Having the equipment and resources in a timely manner to remediate the Manefay and return to full production was one of the keys to Bingham Canyon's success. Without the teamwork between the Procurement Team and a wide variety of vendors, as well as their willingness to go to extraordinary lengths, that critical equipment would not have been ready and the mine would not have been able to accomplish what it did. The Procurement Team and the vendors rose to the occasion and made a tremendous difference after the Manefay.

Lessons Learned After the Manefay

Set stretch targets. The Senior Leadership Team set what was believed to be an impossible target to have operations up and running just two days after the Manefay failure. Without that goal, it could have been several weeks before the mine started getting back on its feet. Enabling the mine to quickly return to work by setting such targets changed the perceptions of employees, Rio Tinto, and the mining community regarding what was possible for the Bingham Canyon Mine.

Empower the workforce. All levels of the company worked together to perform risk assessments, manage change documents, and solve problems after the Manefay. This empowerment helped to motivate the entire workforce and continued to build trust between management and the hourly workforce. This trust is the basis for many of the accomplishments at Bingham Canyon Mine to recover from the Manefay failure.

Set goals, not tasks. The leadership did an excellent job of setting goals for the teams instead of just assigning tasks to be done. This allowed people to be much more innovative in the ways that problems were solved. Examples include the purchase of new equipment and building a new shovel at the bottom of the pit. Much of that innovation stemmed from the company sharing with the employees what the overall goal was and allowing them to solve the problem instead of being given a specific task.

Communication is key. This is probably true for every chapter. The single graphic that broke the remediation work into smaller segments was a great example of the importance of good communication. Without that figure, it would have been much more difficult to communicate the goals that needed to be tackled.

Break down silos. The technical planning became much more integrated with geotechnical, hydrology, and geology as a consequence of the Manefay failure. In addition, the Maintenance and Operation Teams were integrated into the First Ore Team. Both of these are examples of how integration resulted in a better outcome.

Remember that suppliers and vendors are part of the team. The Manefay showed how important it is to have a great working relationship with suppliers and vendors. Price is important, but the value that these groups can bring is often just as important, and in some cases, more so.

Uncovering the Next Ore

Chapter 4

This chapter details the work completed to safely remediate the risks of the head scarp, debris-filled safety benches, and the damaged Bingham Shop while repairing the 10% haul road to mine the next ore body. All of this work had to be completed by the end of 2013, otherwise the exposed ore would be depleted and downstream operations would be halted.

The outcomes of this chapter led to the lessons learned about choosing the right leader and creating a thinking culture so that options are analyzed, bottlenecks are identified and addressed, and tools are fully used. Safety outcomes led to understanding the impact of stepping back to identify potential risks, fully assessing risks before proceeding with a plan to address the risk, and ensuring that there is always a second method of egress.

On April 27, 2013, the First Ore Team reached a remarkable achievement by loading the first truck with ore from the bottom of the Bingham Canyon Mine, just 17 days after the gigantic Manefay landslide. As great as that triumph was, there was no time to sit back and dwell on the accomplishments. The Manefay debris covered approximately half of

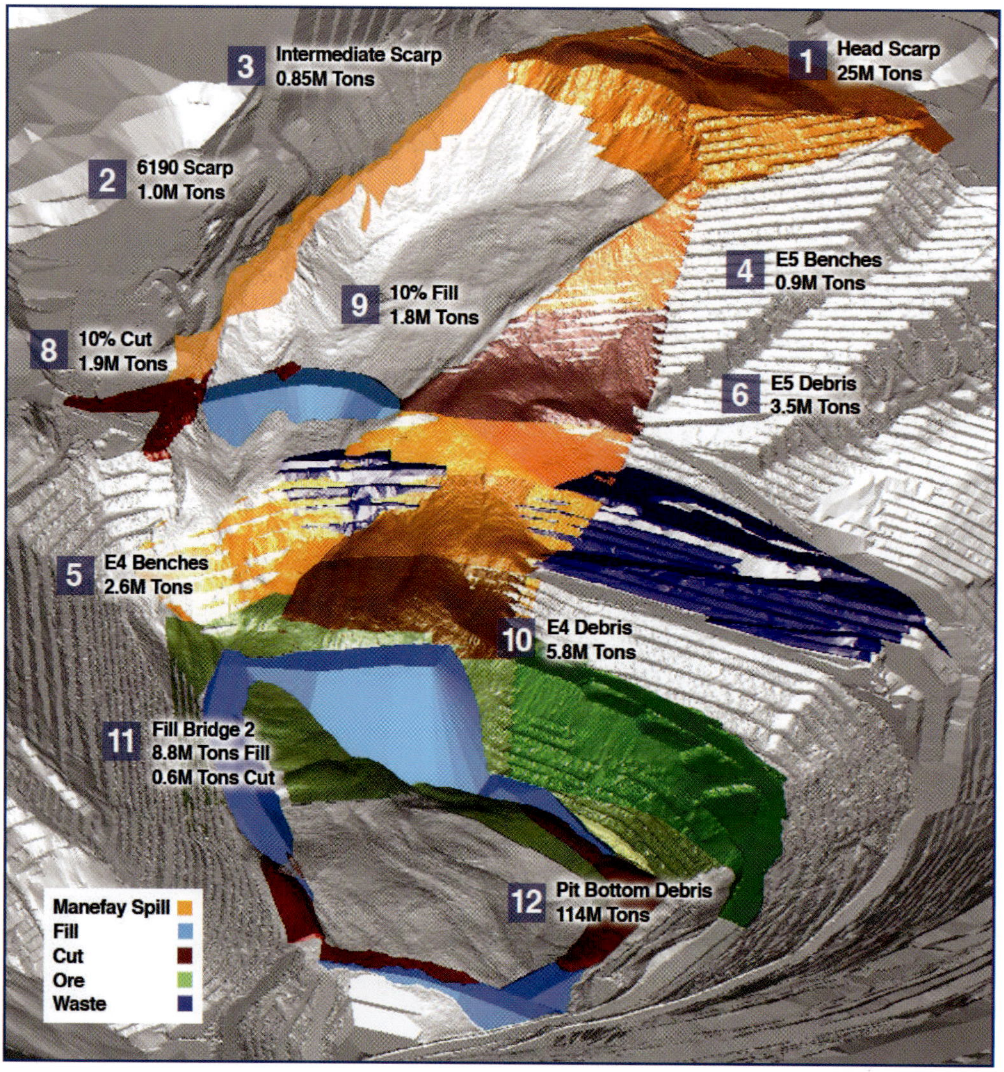

Figure 4.1 • Manefay Impacts

the ore the mine had available in an area called E4 North. The only remaining ore was in E4 South, and at the rate the First Ore Team was mining, they would run out of ore by the end of the year if something was not done soon. Figure 4.1 shows the ore in green. The dark green area was the only remaining ore that could be safely mined just after the Manefay with no additional work. The lighter color green as well as the areas covered in dark brown were ore that was either covered or encumbered by the Manefay debris.

The remnants of the Manefay left several very large hazards that had to be addressed before E4 North ore could be safely uncovered. To prevent Bingham Canyon Mine from running out of ore, it would have to take on what was probably the largest remediation project in the history of mining. To meet this challenge, they created the "Next Ore Team" whose responsibility was to make the Manefay slide areas safe and uncover E4 North before the mine ran out of ore.

The Next Ore Team would have the most challenging job because their work would entail the difficult task of remediating the effects of the Manefay slide, including

- Stabilizing the head scarp,
- Stabilizing the intermediate and side scarps,
- Reestablishing thousands of feet of safety benches that had been filled in with debris,
- Demolishing the Bingham Shop,
- Constructing a new 10% haul road in the slide area,
- Removing debris from the old 10% ramp,
- Removing debris and fill material on E4 North ore, and
- Constructing a new ramp to extract ore.

This work would be highly technical because it entailed the remediation of not only more scarps but larger scarps than had ever been attempted before, and it needed to be done quickly to meet the goal of a continuous supply of ore. The technical part of the work created many uncertainties—the size and magnitude would require new operating methods and equipment to keep people safe as well as get the work done. The new approaches involved not only operational techniques to work on tall scarps, but geotechnical processes to evaluate the stability of the head scarp. Much of the work, at least to get started, would require remote-controlled equipment operations on a scale that had never been done before. This also created significant uncertainty.

The Next Ore Team required a special kind of leader. This person had to be someone technically competent to understand the technology and develop the procedures that would be required. In addition, he or she had to be practical to implement the new technology, motivational to keep the team moving forward when times became difficult as surely they would, and humble to seek advice from others and use it to full advantage. No one had experience with a job of this magnitude. Executing this job effectively would require someone who could leverage the knowledge and capabilities of many people at various levels and not let arrogance prevent him or her from finding a better way of doing things.

On May 2, Matt Lengerich caught up to me at the Rio Tinto Regional Center to discuss candidates for the position of remediation manager. One of the options that Matt had for that position was Cody Sutherlin, superintendent of mine planning. Cody had been doing a tremendous job. He had built up the skills and capabilities of his team before the Manefay, had taken over setting up and managing the team after the Manefay when I was unavailable, and he was key in the integration of numerous engineers from various areas to get them to work together and start to build the future plans for the mine. Although Cody was only 28 years old at the time, he was working as a seasoned veteran and had earned the respect of those around him. In addition to all that he was accomplishing, Cody also listened to and accepted coaching.

My first response to Matt's idea about using Cody as the remediation manager was "No way!" because Cody was too critical to the planning effort, and the planning effort was vital to the future of the company. Matt seemed to accept the explanation and began to consider other alternatives.

That night, I thought a lot about Matt's preference of Cody for remediation manager and why I had been opposed to it. I had actually rejected the idea to make it easier on me. Having Cody in the position of mine planning superintendent made my life much simpler because Cody had been assuming many of my managerial duties. By taking on those obligations, he allowed me to be more effective and focus on higher level issues. In reality, Matt had brought me to Kennecott Utah Copper to mentor future leaders and build capacity in the Technical Services group. Cody was ready to move on to a different position, and I had to concede that it was wrong for me to hold him back. In addition, if Cody moved into this new position, I would be able to mentor his replacement, thus building on my original mandate.

The next morning, I sent a note to Matt stating that I believed Cody was the right person to take on the job as remediation manager. Matt responded that he had several other people to consider, not the least of which was Tim Juvera, the manager of asset management, who had done a great job of leading the First Ore Team.

That evening Matt held a meeting with a few of the managers and Anna Wiley to discuss the selection for remediation manager. I was included in the meeting because I was not being considered for the position although most of the other managers were. As a group, we listed and considered several candidates for the position as well as their pros and cons. Some of the candidates were from the mine, others from other parts of the business. After some discussion, the following criteria were developed:

- Strong leadership
- Understanding of mining operations
- Criticality of current position
- Ability to reach out to others
- Growth potential

Some individuals were not selected because of the importance of their current position. Tim Juvera was one of those. Asset management was seen as a critical job going forward, and Tim was needed to make that part of the business work well given that we would have to do maintenance differently in the bottom of the pit. Mining knowledge was also very important. At the end of the day, there were just a few people that met all of the criteria. The final decision was not made in the meeting. It was late, and Matt decided to contemplate the selections over the weekend. On Monday, May 6th, Matt returned to work and told us he chose Cody to be the remediation manager.

You never know if a decision is the best one until you can look at it in retrospect. In some ways, it was risky to select such a young and still-developing individual as the manager for such a vital position that was hugely critical to the company. But Cody had demonstrated the behaviors that were required to be successful in this position. And later, after contemplating the success and accomplishments of the team, it is clear that he was the right person for the job.

Next Ore Team

Cody's first job as the remediation manager was to build a team and create a plan of work. The team that Cody put together covered a variety of disciplines, backgrounds, and skills but was heavily weighted to operations supervisors that had remediation experience at the mine. The original team included the following members:

- Bruce Sharratt, business improvement specialist
- Don Mallet, operations superintendent
- Craig Fisher, principal engineer, projects
- Simon Mallard, concentrator operations superintendent
- Josh Davis, mining engineer
- James Jung, geotechnical engineer
- Karen Bakken, production support engineer
- Eric Cannon, operational readiness engineer
- Jessica Kozian, safety
- Chad Williams, supervisor
- Gary Brant, operator
- David Tisher, operator

The team also included five supervisors and three senior operators. In addition to these members, Rudy Ganske, a geologist, would join the team later in the summer when the 10% ramp was being constructed. Geoff Bedell also joined the team when blasting commenced on the head scarp.

Each member contributed significantly to the team, but the person that stood out was Don Mallet. Don had worked in the operations at Kennecott for a number of years and was well respected by both management personnel and hourly employees. One of the reasons that Don was so important to the team is because he saw that mentoring was a major part of his job. He worked hard at building the skills and encouraging teamwork, which helped the entire team be more successful.

One of the first decisions Cody made was to use Rio Tinto's Management of Change (MOC) system instead of the sign-off document used by the First Ore Team. Rio Tinto has a computer-based system that maintains the documents and ensures that the MOCs follow a standard format, and it tracks who has approved the MOC. I had worked for Rio Tinto for many years and observed that the MOC system was rarely used effectively. However, the Manefay recovery work changed that, starting with the Next Ore Team. The documents were well written and thought out, and work would not begin until all of the required approvals had been obtained. If there were questions or shortcomings in the MOC, the approvers would not approve the MOC until deficiencies had been addressed. Ultimately the MOC process was an important part of keeping people safe.

One of the first tasks for the team was to create a target completion date for the remediation work. The team had a lot of pressure from the chief operating officer to get the job done as quickly as possible. Cody and Anna Wiley met and decided that the ultimate goal was to have the Next Ore uncovered by Thanksgiving of 2013—an aggressive, and maybe impossible, goal to meet. This was especially true, considering that no one had ever remediated a highwall failure as large and complex as the Manefay before.

The next step was to develop a schedule. Cody met with Elaina Ware and Tim Juvera to inform them of the date that was required to get the work done and to ask for their help in developing a schedule. Elaina facilitated the meeting by asking questions of what had to be done and Tim questioned the timing of the steps. From those discussions, Cody used the map developed by Jon Warner, which showed the work that had to be done, and he added the dates. Considering the unknowns, the schedule turned out to be amazingly accurate in estimating how long it would take to do the physical work. When building the schedule, what was not anticipated was the length of time it would take to complete some of the geotechnical studies on the head scarp, or the number of internal and Mine Safety and Health Administration (MSHA) approvals that would have to be obtained to complete some of the critical work.

Nevertheless, the Next Ore Team was motivated and the Senior Leadership Team gave them both the expectations and resources to get the job done. The next seven months would be the toughest in most of the team members' careers, with high expectations, long hours, hard work, great successes as well as lowly failures, many frustrations, learning of new skills, and the chance to work on a high-performance team. In short, it was the best of times, and the worst.

The Head Scarp

The head scarp stabilization was identified as the bottleneck to all other remediation work. Manned operations could not proceed below the head scarp as long as there was uncertainty about its stability. By May 8, Cody and his team started sharing plans and schedules of the work that needed to be done. They also developed an understanding of what MSHA would require to modify the 103(k) Order so the head scarp could be mined.

The Next Ore Team could not put manned equipment on the head scarp without a change in the 103(k) Order, but this did not mean that they could not start working on the head scarp. The team put together a plan and MOC to use a shovel in an area that had been released by MSHA in an earlier 103(k) Order modification. That area would be a pad on which the shovel could operate safely. Once that work was complete, remote-controlled dozers could be used for pushing dirt from the head scarp area to the shovel that would be in an approved work area. The plan was communicated to MSHA, and they determined that a 103(k) Order modification would not be needed because no humans would be working inside the 103(k) Order limits. On May 12, 2013, the remediation of the Manefay slide started with the first shovel of dirt hoisted from the back of the critical Manefay head scarp (Figure 4.2). This was the first of many milestones for the Next Ore Team.

Figure 4.2 • Starting Work on the Head Scarp

Although starting work on the head scarp was important, modifications to the 103(k) Order would soon be required because the push distance for the dozers to the shovel would become so long as to be impractical. The Next Ore Team started generating a plan for how to operate safely on the head scarp. To accomplish this, there were several meetings held between the Next Ore Team and the geotechnical group to evaluate the best operating practices when working on the head scarp.

The questions that had to be resolved were: Where could the manned equipment operate and where was remote-controlled equipment required? Many variables had to be evaluated:

- If there was a failure on the head scarp, how far back would it propagate?
- If there was going to be a failure, how much warning would there be?
- How long would it take to evacuate people from the hazardous area after they were told to leave the area?

It was important to set a definitive boundary between manned and unmanned work so a clear distinction could be marked and enforced. The initial analysis indicated that 26 million tons of material would have to be removed to make the head scarp stable, and the remote-control work would be slower than manned work. In some ways, the slower work would increase the risk, because it increased the exposure time in hazardous areas.

Ultimately, the Next Ore Team was able to break the head scarp into three distinct areas. The first area was a safe zone that was outside the Manefay bed. In this section, even if the head scarp failed, this zone would be safe because it was not in the mass that would fail. The second area was called the Intermediate Zone. This was on the Manefay bed and a potential failure mass, but it was far enough from the crest of the head scarp that it would not fail in a brittle type of failure where there would be little or no warning. Equipment could be safely operated in this section, as long as the monitoring systems were operational. The area was identified by a line that was a few hundred feet in back of the scarp that would be the limit of failure if the head were to have a brittle-type failure.

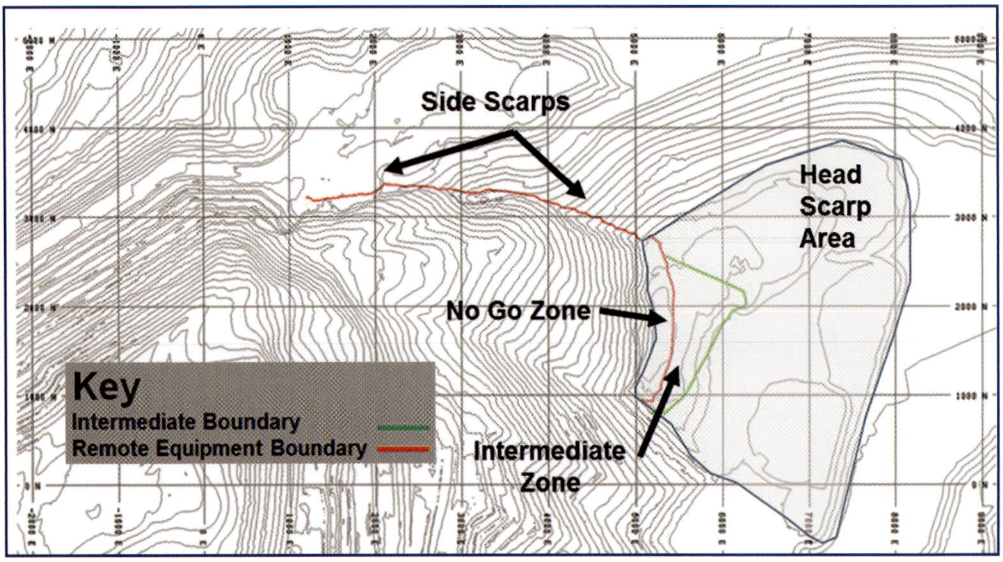

Figure 4.3 • Equipment Boundaries in No Go Zone

On the other side of the line for the Intermediate Zone to the crest of the head scarp was the No Go Zone. In this area, only remote-controlled equipment and operations were allowed. The only exceptions to this rule required a risk assessment, an MOC, and approval from management and MSHA. Figure 4.3 shows the map with the locations of where manned and unmanned equipment could work on the head scarp.

When the Next Ore Team composed the MOC and plan to present to MSHA, they not only showed the No Go Zone, they also provided backup of the geotechnical analysis, a Monitoring and Response Plan, and a time study of how long it would take to evacuate people if an instability should occur. Based on the plan and thorough documentation, the MOC was approved by the company and then MSHA approved the modification to the 103(k) Order on May 20, 2013.

During this period, the Next Ore Team received support from Zip Zavodni and Martyn Robotham. Martyn was very familiar with the conditions and people at the mine. Just after the Manefay failure, Matt Lengerich asked Martyn to come out and give the mine a hand. The combination of Zip and Martyn provided world-class support in the geotechnical field. Zip is recognized as one of the top geotechnical experts in the industry and has a long history with the mine, having done research on the mine as far back as the 1970s.

Martyn helped the mine for more than a month right after the Manefay failed and also returned for more visits as the year progressed. He often played the role of communicator and facilitator—making sure the geotechnical engineers knew the priorities, and the mining engineers as well as management understood the issues facing the geotechnical engineers. With Martyn's help, everyone was better informed and the process evolved.

Lean Board Meetings

Soon after the Next Ore Team was formed, they started using a business improvement tool called "Lean Board" meetings to discuss the results from the previous day and the plans for the current day. Lean Boards were used at Kennecott before and after the Manefay to varying levels of success. In this case, Bruce Sharratt, a business improvement specialist, helped the team modify the meeting format for the Next Ore Team. As with most Lean Board meetings, attendees had to stand during the meeting. The meeting was focused around a set of charts and tables that were manually updated just before the meeting, and because there were no chairs, participants tended to finish quickly instead of standing in one place for long periods. As with all meetings at the Bingham Canyon Mine, this one started with a discussion about safety and a Zero Harm share. After the safety share, the team reviewed the previous day's performance and then ended with a list of critical items that needed to be completed, who was responsible for them, and when they would be completed.

The Lean Board meetings were extremely effective for the Next Ore Team. The team continually changed the graphs and data because of changing priorities as tasks were completed and others started. For example, at one point, the number of remote-control systems that needed to be installed was a bottleneck, so charts and tasks were focused on getting the systems installed. Another time, the utilization of dozers was a key issue to uncover the north side of the E4 cut, so a different set of graphs and tasks became the focus for the meeting. Figure 4.4 shows Cody Sutherlin updating one of the graphs during the daily Lean Board meeting.

Figure 4.4 • Lean Board

The Next Ore Team started every morning with a Lean Board meeting so everyone on the team knew where they were supposed to be and what the team's priorities were. Usually 10 to 15 people attended the meetings. Tuesdays were different, though, because people from outside the Next Ore Team were invited to attend. The Tuesday meetings were crowded, with 30 or more present. The additional attendees usually included the Kennecott managing director, vice presidents, general manager of business improvement, the mine's general manager, operations manager, technical service manager, maintenance manager, and a variety of superintendents and supervisors. This was an excellent meeting for keeping everyone up to date regarding the team's progress and priorities, which were, in reality, the priorities of the entire company, because Kennecott relied on the success of the Next Ore Team.

The large weekly meetings conducted from May into June went well. The Next Ore Team was making progress in removing material and reaching their targets. But in early June, progress started to slow. The Next Ore Team had mined most of the easy waste material and was not able to blast the hard in-place rock. As progress diminished, the team started to fall behind schedule. At that point, it seemed that the senior leaders who attended the meeting each week would have more questions and make more comments. Although these queries and statements were made with the best of intentions to show support for the team and offer advice and guidance, they soon started having a negative effect. The meeting started taking longer than the scheduled 30 minutes. Worse, it seemed like Cody, the Next Ore manager, was losing control of the meeting as well as the confidence of the Senior Leadership Team. By the end of June, the weekly meetings were having an adverse effect on the morale of the Next Ore Team.

There were many reasons for the slowdown. Internal and MSHA approvals were required to start blasting the head scarp, and the easy mining of waste material had been nearly mined out. Equipment had been ordered but was still being delivered, and importantly, much of the remote-controlled equipment had not been commissioned. Operators were still being trained because the remediation work required many more skilled dozer operators with remote-control operating skills than the mine had available. The Next Ore Team was in a planning-and-building phase that was critical to the ultimate success of the project, but not much visual progress was being made in physical remediation work. This lack of advancement as well as the overly optimistic early projections frustrated both the Senior Leadership Team and the Next Ore Team.

Significant work was progressing, but it seemed invisible. It can be difficult to notice new plans, approvals, risk assessments, and training when the goal is to move dirt off the head scarp. There was some physical progress going on at this time. The Bingham Shop was being demolished, and the team was building access and infrastructure so that work could commence on remediating the intermediate scarps. This preparatory work always takes more time and effort than anticipated, and schedules do not adequately reflect that progress. But it was this time-consuming preparation work that would ultimately make the remediation effort much more successful just a few weeks later.

Cody had to take back control of the weekly Lean Board meeting if he was to rebuild the morale of the team as well as maintain his own confidence. Resuming control of the meeting may sound easy, except that Cody was a 28-year-old newly minted manager and the people he had to control were general managers and the managing director of the company. What Cody did was to individually meet with the managing director and general managers to discuss the issue. Cody offered to step aside if they had lost confidence in him. Conversely, if they did not want him to step aside, the managing director and general managers would have to change their behaviors in the weekly meeting. From that point, Cody required them to limit their questions, comments, guidance, and advice during the meetings. They would be able to ask Cody questions and provide comments, guidance, and advice outside the meeting, preferably in private.

Fortunately, the senior leaders did not want Cody to step aside and therefore agreed to act differently during the meetings. Starting the very next week, the entire tone of the meeting changed. Cody was obviously in charge once again, and the meeting ended on time. The morale of the team quickly changed as well. The grumbling and comments that were being made just a short time before seemed to go away and people appeared to be more energized. Perhaps this was because some of the bottlenecks were broken down with the delivery of a drill and the start of blasting, or maybe progress was made because attitudes were changed. Whichever the case, in just a few weeks the team and schedule were back on track.

The atmosphere of the Lean Board meetings continued to improve whenever the Next Ore Team assembled. It was the first time that I had witnessed very effective Lean Board type meetings and their value when they are well managed. Observing the transformations that took place in those meetings changed the way that I subsequently managed my Lean Board meetings with my own team, and the result was better understanding and teamwork.

Mining the Head Scarp

After the 103(k) Order modification on May 20, the Next Ore Team started mining more of the head scarp. The head scarp was situated with a large piece of intact rock sitting on the Manefay bed that was exposed to the surface on the south side of the scarp. This rock was very hard and difficult, if not impossible, to mine if it was not blasted. On the north side of the head scarp was a large area of waste that had been placed there decades earlier from previous mining. Figure 4.5 shows the layers of rock and waste material that made up the head scarp. The buff-colored material on the left side of the photo is waste material from old mining dumps. The darker material is in-place rock.

Figure 4.5 • Head Scarp Material

The May 20th 103(k) Order modification did not allow people into the No Go Zone, so the Next Ore Team had no method of blasting the intact rock. The team tried to rip the hard rock with remote-controlled dozers but found out very quickly that the dozers could not rip or break the intact rock. The dozer ripper or blade would just bounce off the rock. Technology was available to drill the blastholes with remote (operated by a person not on the machine) and automatous (equipment operated by a predefined computer program) drills, but those drills were on order and would not arrive until late May. However, blasting could only be done manually because there was not a method for remote-controlled machines to load a blasthole and tie in the blasting caps and primers to set off the blast.

There was significant pressure on the Next Ore Team to keep up production to complete the remediation work as soon as possible. The First Ore Team continued to increase their daily production rates, which increased the necessity to uncover the north side of E4 by Thanksgiving, in less than six months. With such a short window, every day counted. To keep production up, the Next Ore Team mined as much of the previously mined stockpiled waste material as quickly as feasible. It was easy pushing for the remote dozers and trouble-free digging for the shovels, so the production rate was good—even higher than expected. But there was a problem with this easy digging; so much waste material was removed that what remained was a small patch of intact rock.

Figure 4.6 • Drill Working on Head Scarp

Usually, a small area of rock would not be a problem, but in this case, it became an issue because the area was rough to begin with and the remote-controlled dozers could not create a level pad for the remote-controlled drills to work on. The flat areas, or pads, for the equipment to work on are called benches and are important for the equipment to work in stable and productive conditions. Kennecott typically removes 50 feet of vertical material at a time in the building of a new bench. However, the mining of the head scarp would be done in much shorter 20-foot lifts. Until the first bench could be established with blasting and mining, it created a situation where the mining of the head scarp was a slow process. Even after the first bench was mined, it was slower than normal operations because of periods when the shovels had to wait for blasts and the monitoring after the blast to ensure there were no geotechnical instabilities before starting to mine. Figure 4.6 shows an autonomous drill working on uneven ground on the head scarp.

This was a hard lesson for the Next Ore Team—rushing to show production and gains in the short term can slow the long-term performance of the project. This happened not only because of the pressure to produce from senior leadership, but also because people wanted to see progress and feel successful.

Remediating the head scarp created a lot of concern and discussion for the Next Ore Team and the company's management. One option was to blast the head scarp with enough explosives that it would completely fail and slide into the pit. The second option was to mine the top of the head scarp in lifts similar to a normal mining operation.

The first option appeared to be the quick answer: start drilling numerous blastholes on the head scarp and load as much explosives as possible to recreate a Manefay-type failure. The pit would be evacuated and the material could be mined at the bottom of the pit. What could be easier? But after more detailed analysis and risk assessments, this method had serious problems. The first dilemma was how to drill the entire head scarp area. The drills would not have enough level working area to drill a majority of the head scarp. Several benches would have to be mined using traditional mining methods just to provide appropriate drilling conditions. The second issue was that the blastholes would have to be several hundred feet deep. This type of drilling is done for blast casting in coal mines for dragline operations, but acquiring the appropriate drills and drilling the deep holes in hard rock for the entire head scarp would take several months. However, the biggest problem with this option was the possibility that the head scarp would not completely fail and consequently leave an unstable mass. This mass would require remote-controlled equipment for all remaining work and would also be difficult to monitor and thus continue to be an even greater danger. Furthermore, it could take longer than a year to fully remediate the scarp, possibly creating a gap in ore production.

The second option was to mine the scarp in benches using traditional mining methods. This would require blasting just one bench at a time and moving the material with trucks and shovels, which at first seemed to be the slower and more expensive route. However, with this method there would be a point where a reasonable factor of safety could be attained and manned operation initiated below the head scarp. After studying both options in detail, a decision was made to mine the head scarp in the traditional manner.

First Blast

Blasting in the No Go Zone of the head scarp had become the critical path item in remediating the Manefay. It required drilling in difficult conditions with either a remote or autonomous drilling rig, neither of which was currently at the Bingham Canyon Mine. By June 13, Flanders Electric installed one of its Ardvarc systems that can be used to operate drills both remotely and autonomously. This system was installed on a DML blasthole drill that was already on-site.

With this system, the blastholes could be drilled without having an operator on the drill while it was in the No Go Zone. Installation and commissioning a remote-control system, especially a retrofit on an existing drill, is a complex process that usually takes months of planning and implementation. The mine did not have that kind of time, and subsequently the system was quickly installed. However, learning to drill with a remote-control system on the head scarp and overcoming commissioning bugs proved to be a difficult process. This was not only because of the complexity of the system, but also because the ground where the drill needed to operate was very uneven and rocky as a result of trying to prepare the area with a remote-controlled dozer in extremely hard rock. If the team had not mined the entire waste stockpile in that area, it would have been easier to use that material to level off the bench so the drill could work efficiently. Figure 4.7 shows the DML drill being controlled remotely to bore some of the first holes in the No Go Zone of the head scarp. Notice how far the right-hand track is off the ground because of the rough terrain.

Figure 4.7 • Drilling on the Head Scarp

From early to late June, drilling progress was slow because of the problems the team was encountering. Stephane Leblanc, the chief operating officer, was pushing the team to get the drilling done without delay. If the team needed more drills, they had the authority—even the expectation—to hastily acquire additional drills. As Stephane told the team, "If the drill you need is in Australia, hire the largest plane in the world and get it here tomorrow." Stephane knew that eliminating the head scarp bottleneck was the key to preventing a large gap in ore production several months later. The economic impact of the ore gap would be much larger than the cost of getting whatever equipment was needed to get the job done.

Everyone was aware that the drilling work was the bottleneck. Flanders (the remote-control machinery manufacturer), the remediation team, and a contractor (Mark Sauder) worked feverishly to resolve the commissioning issues and develop the procedures necessary to work remotely on uneven terrain. These were difficult problems that could have taken months to work out, but the team had the drill capable of drilling the holes on the top of the head scarp by June 25. At the time, it seemed like it took a long time to fix the issues. In retrospect, it was a tremendous accomplishment that the team was able to install a new system on an existing drill and get it to operate in such difficult conditions in less than a month.

In the previous weeks, blasting experts, such as John Floyd from Blast Dynamics and Colin Matheson of Matheson Mining Consultants, were brought in to work with Geoff Bedell, the mine's blasting engineer, as well as the company's geologists and geotechnical engineers to develop the blasting methods for the head scarp. A blast was needed to break the rock, but not so much that it would start a landslide or create unstable ground. The solution was to perform a small nine-hole test shot on the head scarp to measure the scarp's response to blasting. Results of the test shot would then be used to help design the future shots. Plans for extensive monitoring of the head scarp were made before the blast. Figure 4.8 shows the relation between the sizes of the blasting area compared to the amount of monitoring. The area of solid rock was only 600 feet by 200 feet (in light blue). The dark blue area is the head scarp. TDR (time domain reflectometry) holes were drilled to monitor movement on the base of the head scarp that rested on the Manefay bed. Multiple seismographs were stationed to measure ground vibration. Four different radars would monitor the blast to check for any ground movement that would indicate a potential highwall failure. Three microseismic drill holes were installed as well as a downhole extensometer. New prisms were installed that were read with robotic theodolites, plus a new GPS (Global Positioning System) prism unit was purchased so three-dimensional (3D) movement could be monitored.

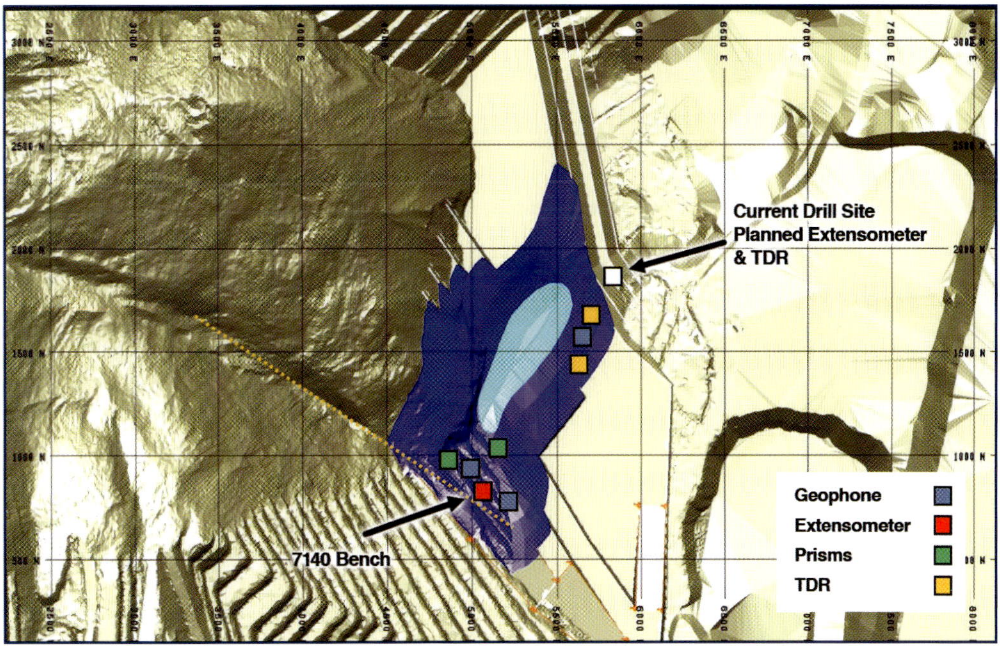

Figure 4.8 • Blast Monitoring

Plans were also made to evacuate the mine before the blast. People would not be allowed in the bottom of the pit for at least two hours after the blast to provide a stand-down period to monitor movements on the head scarp and ensure that the blast did not create instability.

Although the remote-controlled drilling could be performed without people in the No Go Zone, blasting required that humans enter the zone to load the blastholes and tie in the shot. Before the blasters could work in the No Go Zone, another detailed blasting plan, plus a detailed Semi-Quantitative Risk Assessment, and an MOC had to be completed and signed off. This planning process included a team of independent blasting experts, geotechnical experts, internal blasters, engineers, and safety professionals.

The team developed a plan where the blasters would enter the No Go Zone wearing fall protection or lifelines that were tied to an MSHA-approved anchor that was located outside the No Go Zone. Consequently, if a failure did occur while the blasters were inside the zone, they could be pulled back to safety. All work had to be performed manually, because equipment was not allowed to be operated inside or near the No Go Zone when the blasters were in the area. Geotechnical engineers had to actively observe the monitoring systems so that blasters could be quickly evacuated if there was any indication of potential highwall failure in the head scarp. Figure 4.9 shows the Next Ore Team testing the blasting procedures before submitting the plan to be approved.

Figure 4.9 • Testing the Blasting Procedures

The blasters practiced loading holes using the set before starting to work in the No Go Zone. The work proved to be difficult and slow. Blasting agents were loaded into a wheelbarrow and manually shuttled to the blastholes. Figure 4.10 shows the blaster tying in the holes for the first test blast on the head scarp.

Although the use of lifelines worked for small detonations like the test blast, as the number of holes and amount of explosives increased, the method became problematic. The lifelines were easily tangled and added additional work given that the ropes had to be pulled along and could be caught between rocks. The use of the lifelines more than doubled the time to complete the job of loading and tying in the blastholes. Although this approach was used for several blasts, a later Semi-Quantitative Risk Assessment showed that the additional time spent on the blast pattern actually increased the risk to the blasters, so the use of the lifelines was discontinued.

Figure 4.10 • Loading Blastholes for the Test Blast

By June 25, the test drill holes had been drilled using new autonomous and remote-control technology, and plans were in place for blasters to work in the No Go Zone. In addition, the blast design was completed, the monitoring plan was in place, the MOC approvals had been signed off, the Mine Technical Review Team (a group of Rio Tinto and independent experts, described in chapter 7) supported the work, and MSHA had agreed to modify the 103(k) Order to allow test blasts on the head scarp. Never before had as much time, energy, review, approval, and cost been put into one simple nine-hole blast in the history of Bingham Canyon Mine (and perhaps the entire mining industry). But there had never before been a blast more important to a company. The nine holes were in reality a test shot to understand the characteristics of the rock and demonstrate that the blasting could be controlled to prevent a failure on the head scarp. This one blast would be the start of the Bingham Canyon Mine being able to mine the intact rock in the head scarp that held up much of the remediation work. If this shot could not be safely drilled, loaded, and controlled with this method, the team would have to start over and the remediation could be delayed for months.

On June 26, everything and everyone was ready. Part of the plan called for a geotechnical engineer (James Jung) to join the blaster, Fred Hazelwood, as the blastholes were loaded in the No Go Zone. The engineer would monitor the conditions on the ground and call for an evacuation if he observed any potential hazards. Figure 4.11 shows James monitoring the area as Fred prepares to load the holes. After the blaster loaded and tied in the nine holes, he and the geotechnical engineer moved to a safe distance from the blast. The pit had already been evacuated and the blaster prepared to detonate the shot. In a matter of milliseconds, the work of the previous four weeks came to bear as the explosion detonated. Since the nine holes were only a test shot, the holes were relatively shallow at only 20 feet deep, with a low powder factor of 0.28 pounds of explosive per ton of rock. The goal was to limit the amount of vibration, called the peak particle velocity, or PPV, to below 0.2 inch per second on the Manefay bed. Keeping the PPV low would prevent damage to the bed and minimize the risk of a highwall failure.

Figure 4.11 • Blaster and Geotechnical Engineer on the Head Scarp

Figure 4.12 • First Blast on the Head Scarp

The blast went off just as planned and the PPV remained below 0.2 inch per second on the Manefay bed. The team demonstrated they could safely drill and load the holes and control the PPV. The test was a complete success. The PPVs were almost exactly as predicted and the team had shown they could blast and protect the head scarp from failing. Now it was time to start the production blasting. Figure 4.12 shows the test blast as it was going off.

The first production blast was planned to be only 50 holes, but the depth was increased to 30 feet and the powder factor increased to 0.30 pounds of explosive per ton of ore. This was still conservative based on the results of the test blast. The PPV would be low on the first bench of mining the head scarp because of the great distance from the surface to the Manefay bed. Each succeeding bench would get closer to the bed, so the PPV would be increased. The designs called for keeping the PPV on the Manefay bed below 5 inches per second until the last benches where the drill holes were right on top of the bed.

The plan for performing the production shots required an MOC as well as Mine Technical Review Team and MSHA approval, just like the test. These approvals were critical to the timeline, and MSHA was always the last group to approve any plan (by design, because they had ultimate authority under the 103(k) Order). Having a good working relationship with MSHA was extremely important at this point. The company kept the local and district MSHA officials informed as to what plans would be submitted as well as the critical nature of the submittals. The MSHA inspectors were at the mine on a regular basis (sometimes daily). Consequently, they also had a very good handle on the work being done and were assured that the plans were being followed.

It was almost the first of July before the plan for the production blasting was submitted to MSHA. There was a lot of concern that the remediation would be delayed because it was the start of the Fourth of July holiday and many people took the week off for vacation, especially in the government sector. Senior MSHA leadership officials had made a commitment to try and review plans for changes to the 103(k) Order within a week's time because they understood the importance of the remediation work. Getting the appropriate people to review the plans during a holiday week was going to be a significant test of that commitment. Up until July, the MSHA inspectors and technical experts who reviewed the plans had been responsive to the time requirements. They checked the plans thoroughly and, up to that time, were able to reply in a timely manner. At that point, however, every day of delay in the approval would be a day's delay in the getting the mine back to full production. Being Fourth of July week, there was a risk of not only one day's delay, but potentially a few weeks of delay as people went on their summer vacations.

To MSHA's credit, it stood by its commitment. MSHA reviewed the plans both in the local office as well as the technical office in Pennsylvania and made the required modification to the 103(k) Order on Wednesday, July 3rd. MSHA's willingness to work with the company to respond to the plans for changes to the 103(k) Order made a difference in the remediation and was a great example of how important it was to have the support of external groups. Kennecott treated the MSHA approval process as a key step in helping to keep people safe. In reality, the work and plans that were reviewed by MSHA needed to be done for our internal process as well, but MSHA was one more independent review that might find something we missed. This was a good final check before the company proceeded with work that it had never performed before. In return, MSHA understood our processes, level of planning, and focus on keeping people safe, which helped build a high level of credibility. The outcome was probably the best example of an effective relationship between a government regulatory enforcement agency and a company that resulted in keeping people safe and the company in business.

The Fourth of July was celebrated at the Bingham Canyon Mine with the first production blast on the head scarp. The extensive geotechnical monitoring was in place and the blast went off without incident. A few hours later, the remote-controlled dozers were pushing dirt to the shovels, and production had started on removing the intact rock above the Manefay bed.

Bingham Shop

The head scarp was the key remediation work after the Manefay failure and the critical path for preventing a gap in ore production. Until the head scarp was made safe, work could not proceed on critical tasks, such as rebuilding the 10% haul road that had been destroyed or uncovering the ore that had been covered up by the debris.

Although the first priority for the Next Ore Team in June was to start blasting and mining the intact rock on the head scarp, there were several other ongoing projects that needed attention. They were not necessarily the critical path items to the schedule at the time, but if left unattended they could become a bottleneck. These projects included sloping the intermediate scarp and 6190 scarp to the west of the head scarp, cleaning the E5 benches below the head scarp, and demolishing the Bingham Shop. Each of these projects had its own unique issues and challenges, and needed to be completed in a timely manner, otherwise future work would have to wait.

As can be seen in the earlier photos in Chapter 1, the Bingham Shop was severely damaged by the Manefay and there was no hope of repairing the building. It was precariously hanging over the far west scarp and therefore was in the No Go Zone for using manned equipment.

Eric Cannon was in charge of safely demolishing the shop, but he needed to do it quickly enough to ensure that the demolition would not impede the construction of the new 10% ramp that would be built where the building currently stood. The shop would have to be demolished before the head scarp was remediated. This meant the Bingham Shop would have to be demolished with remote-controlled equipment.

Eric was not able to find a company that had ever demolished a building remotely. In fact, the building could not be physically inspected before the job began because it was in the No Go Zone. Through some searching, he did find a company willing to give it a try if the Bingham Canyon Mine supplied a remote-controlled excavator.

By June 4, the demolition crew was ready to start work. Approvals to demolish the shop had been granted by MSHA shortly after the Manefay slide. A berm was constructed to delineate the No Go Zone and people were not allowed beyond that point. Considering the extensive structural damage to the building and the way the edge of it hung out over the scarp, it was easy to convince people not to enter the area.

The demolition proceeded with the remote operator standing behind the No Go Zone berm to operate the excavator. The excavator was fitted with large hydraulic jaws that can cut through steel beams. These jaws were used to cut critical structures and then the excavator would pull or push down the structures from as high up in the building as it could reach and then worked its way down to the floor. When there was enough building debris on the ground, the operator used the excavator to drag the debris to a safe area outside the No Go Zone. The material was then sorted, and the recyclable material was shipped to a recycling center and waste material taken to a solid-waste center. Figure 4.13 shows the remote-controlled excavator taking down the main portion of the Bingham Shop.

Figure 4.13 • Demolition of the Bingham Shop

As with many other Next Ore projects, the Bingham Shop demolition was a learning experience. The initial work on the shorter end of the building went quickly. When work started on the main part of the shop, which was much taller, the progress slowed considerably. But Eric and his team continued to make headway, and by July 8, the Bingham Shop had been demolished and the building material was removed from the site. The demolition of the Bingham Shop was done well before the completion of the head scarp stabilization and did not become a bottleneck.

The Dangling Dozer

Not all of the remediation work went according to plan. On June 10, one of the most challenging incidents for the Next Ore Team occurred when a D10T remote-controlled dozer went over the crest of the scarp. The dozer was pushing material in the No Go Zone in the head scarp to the shovel that was loading trucks outside of the zone. The remote-control operator was almost 1,500 feet away and he could not see the back side of the dozer. The operator backed the dozer onto the crest berm of the scarp and the crest gave way, dropping the dozer partially over the edge. Figure 4.14 shows the dozer hanging precariously on the top of the nearly 600-foot scarp.

Fortunately, the dozer did not tumble down to the bottom of the head scarp. When the crest broke away, the dozer slipped into the resulting channel over the edge and came to rest with the dozer ripper wedged into the rock. It then sat with the blade facing the top edge of the crest. No person was at risk in the incident because the dozer was being operated remotely, but given that the dozer was valued at more than a million dollars, the company did not want to lose it. The steep angle of the dozer prevented it from being able to crawl back to the top of the crest. If workers had rushed to free the dozer by trying to remotely have the dozer try and crawl up the slope, there was a good chance that the dozer would then slip and tumble down to the bottom of scarp and the piece of equipment would have been lost.

This event had occurred in the late morning. It would have been easy for the team to make a quick plan and rush into retrieving the dozer, but Matt Lengerich directed them to complete a full risk assessment so that the dozer could be recovered with no additional risk to people or equipment. Photographs were taken of the dozer position so the full situation could be carefully evaluated. The team worked late into the night and developed a plan to recover the dozer by the next morning. Before performing the actual recovery, the Next Ore Team walked through all the steps that needed to happen, so they were prepared for recovering the dozer.

Figure 4.14 • D10 Dozer on Head Scarp Wall

The plan was to use a 390 Caterpillar remote-controlled excavator to dig a trench below the blade of the dozer. The excavator would then be tethered with a one-million-pound tow rope and secured to a D11T dozer to prevent the excavator from being pulled over the scarp if the hanging dozer should fall during the recovery. The excavator was then positioned so the bucket could reach behind the blade of the dozer to help pull it up and over the crest. The hanging dozer was still operational and its power could be used to help move itself during the recovery. The day before, the team had determined that the dozer would be shut down during the night so it would not run out of fuel, but periodically restarted and idled to keep the battery fully charged. The plan then called for three pieces of equipment working together but all being controlled remotely—not an easy task.

Figure 4.15 shows the million-pound tow rope that was hooked up between the remote-controlled excavator and the remote-controlled D11T dozer.

Senior leadership was notified of the incident and supported the recovery plan. None of the senior leaders wanted to lose a new dozer, but from the very beginning of the remediation efforts, everyone knew that there was a possibility that some equipment could be lost. After all, the reason that remote-controlled equipment was being used was because of this increased risk. The important part was that people's lives were not put in jeopardy.

Later on the morning of June 11, the team was ready to implement their plan. The best remote-control operators were selected to operate the excavator and dozers. Just like the steps in the plan walk-through, the excavator was moved into place and a trench was dug below the dozer blade until the top of the dozer tracks could be seen. The excavator was then brought back outside the No Go Zone and the million-pound tow rope was attached. The rope was anchored by a D11 dozer that was ready to help pull the excavator if needed. The excavator was then moved back into position in front of the dozer, and the arm of the excavator was extended so the teeth of bucket were on the back side of the dozer blade. When the bucket was in position, both the dozer operator and excavator operator worked together to increase their throttles to ease the dozer up and over the crest of the head scarp. Figure 4.16 shows the dozer and excavator working together to extract the stranded dozer from the crest of the head scarp.

Figure 4.15 • Million-Pound Tow Rope

The dozer recovery worked almost exactly as planned. The only damage to the dozer was a small dent in the rock guard on the top of the blade. After an inspection by the Maintenance group, the dozer was put back into service. More importantly, no people were put at risk in completing the task. No one had to enter the No Go Zone, and the dozer was recovered quickly and efficiently.

Figure 4.16 • Recovering the Stranded Dozer

What started as an unexpected incident that could have been a serious setback to the Next Ore Team actually turned out to be an exercise in teamwork and good planning. The Next Ore Team came through this test with more confidence in themselves and their fellow team members plus they became even more committed to thinking through the task before undertaking something they had never done before.

E5 Safety Bench Cleaning

Large open pit mines like the Bingham Canyon Mine use safety benches to prevent rocks from rolling from the top of the mine and not stopping until they reach the bottom. The safety benches catch rocks that are rolling or falling down the highwall and are therefore a critical safety device. Material from the Manefay failure fell out and into the pit as the mass began to collapse. This material filled in or covered the safety benches from the 6990 bench elevation that was near the top of the Manefay mass, down to the 6240 bench level, which was just above where the new 10% haul road would be constructed. The work would require that 16 safety benches be reestablished in just a few months when the 10% would need to be built to start additional mining operations. To make the task even more difficult, much of the work would have to be done with remote-controlled equipment, at least until the head scarp was stable enough for manned equipment to be operated in the No Go Zone. MSHA had approved the modification to the 103(k) Order on June 26 so that work could be started on the E5 bench cleaning with both remote-controlled and manned equipment. Figure 4.17 shows the Bingham Canyon Mine after the Manefay with an inset showing the location of the E5 benches that had to be repaired to make the 10% haul road safe.

Figure 4.17 • E5 Safety Benches

Cleaning of the E5 safety benches started at the very top bench and worked down to the 10% haul road. Manned excavators and dozers were used outside the No Go Zone, and remote-controlled equipment was used inside the zone. The safety benches were too narrow to load debris into a truck to be hauled away. Material was either pushed or cast off the bench being cleaned down to the next lower bench. The material would slide or roll down the highwall until it reached the angle of repose or was caught on a flat area and came to rest. Some of the Manefay debris had to be rehandled on every bench. It also required the team to plan the operations so that when multiple benches were being cleaned at one time, the equipment on the upper bench did not push debris and damage equipment on the bench below (Figure 4.18).

Figure 4.18 • E5 Safety Benches Being Cleaned

One of the concerns with uncovering the benches was that the Manefay debris may not have only filled in the safety bench, but the force and weight might have broken or sheared off the front of the bench, making it ineffective for capturing falling rock. Although this was not a large issue in the E5 area, it did become a problem later on the west side of the pit where the Manefay hit the wall of the pit and left the high-water mark, as discussed in Chapter 1.

Remote-controlled dozers had been used to clean safety benches at the mine in years past, but the E5 remediation was a much larger volume of material and it was the first time that remote-controlled excavators were used. One of the difficulties for the remote-control operations was the lack of visibility. The equipment would begin its task in close proximity to the operators, so they had a good line of sight, but as the equipment worked along the bench and further from the operator, visibility became more difficult and the operation suffered. One of the solutions was to install cameras on the equipment so the operator could control the equipment remotely while watching through

Figure 4.19 • Remote Dozer Operation

a video monitor. This technology was installed and used for building a small road to tie into the 10% haul road. The Next Ore Team had created another tool to help with the E5 bench cleaning.

By August 1, the Next Ore Team had finished cleaning the eighth bench. The cleaning was almost half complete. Simon Mallard and his team were well on their way to having the E5 safety benches cleaned in time to keep the new 10% haul road safe from rock falls from the Manefay debris.

Intermediate Scarp

By mid-July, progress was being made on the head scarp, and the Bingham Shop demolition had been completed. Remediation was started on the next set of projects, the intermediate slopes and the 6190 knob. Work on the intermediate scarps was fairly straightforward because they were just over-steepened slopes that were primarily in previously stockpiled waste material. The problem was that some portions of the scarps were nearly 600 feet tall and the material was not competent.

Given enough time, the intermediate scarps would naturally slough material off of the over-steepened scarps, which would result in a natural slope that was at an angle of repose and stable. However, the concern was that if a large portion of the intermediate scarp were to fail at once, then the debris could reach the 10% haul road that was soon to be constructed, creating a hazard to people and equipment.

The plans for remediating the intermediate scarp called for using remote-controlled excavators and dozers to cut down the front edge of the scarp until an overall slope at an angle of natural repose (or lower) was developed. Once the angle of repose was achieved, manned dozers could be used to continue pushing downslope. Figure 4.19 shows an operator on the ridge (upper right) remotely controlling a dozer.

Work on the intermediate scarp started at the top near the head scarp and progressed downslope until it tied in to the 6190 knob area. Work progressed well on the scarp with only one minor incident. On July 27, a remote-controlled dozer operator was distracted by a loud cracking sound when another dozer hooked a piece of wire connected to a power pole. The dozer was allowed to travel further down the slope and was not able to back up over a small ridge. Again, a full risk assessment and plan were developed but a different approach was taken this time. Since the scarp had already been pushed down to the angle of repose, it was decided that the million- pound tow cable could be safely attached to the back of the ripper by a person. After that was done, the stuck dozer was pulled out with a D11T dozer.

The 6190 Knob

The 6190 knob remediation started at approximately the same time as the intermediate scarp remediation in mid-July. The 6190 knob was also an over-steepened scarp that was the result of the Manefay failure, but unlike the intermediate scarp, the knob was in solid rock instead of stockpiled waste material. Remediation of the 6190 knob was important because it stood above the 6190 Complex and the future 10% ramp. Both were critical infrastructures for the mine and were threatened by a possible failure of the 6190 knob (Figure 4.20).

The remediation of the 6190 knob would be similar to the head scarp treatment. The work on the 6190 knob was covered under the same 103(k) Order revisions as the initial blast and production blast revisions for the head scarp because the same blasting procedure and distances for the No Go Zone would be observed. But in some ways the 6190 was more difficult because the ridge and rock knob were very narrow, limiting the working room for the remote-control operators and equipment.

Figure 4.20 • 6190 Knob

In preparation for the Manefay, drilling had been started to improve the monitoring of movement of the Manefay bed. The drillers had been quickly evacuated from the area the night of the Manefay failure, leaving the equipment behind. Before work could begin on the 6190 knob, an additional 103(k) Order would be required to move the drilling equipment so it would not be destroyed by rocks falling from the remediation work. Figure 4.20 shows the 6190 knob of rock above the contractor drill and associated equipment on the road that once led to the Visitors Center.

Simon Mallard was responsible for the 6190 knob remediation. The actual recovery of the equipment was relatively fast and easy. Because it was on a roadway, the debris that had rolled down from the 6190 knob was easily pushed aside with a dozer. Spotters and geotechnical monitoring equipment were placed so that the equipment could be removed safely. In a short time, equipment was either driven out or loaded and hauled away.

The remediation of the scarp was not so easy. Work began with one remote-controlled dozer building a single track access across the top of the ridge going to the top of the knob (Figure 4.21). The knob itself was hard yet brittle rock and nearly vertical. Because the area was so narrow, there was a risk that a piece of rock would fail and take the piece of equipment with it. Of all the remediation work that had to be done, the Next Ore Team thought this was the area where they were most likely to lose a piece of equipment.

Figure 4.2 • Dozer on the Ridge of the 6190 Knob

Figure 4.22 • Drilling and Dozing on the 6190 Knob

Once the trail had been built on top of the 6190 knob, several methods were tried to break and move the hard rock. The rippers on the dozers would not penetrate the rock, and the teeth of the excavators would simply bounce off it. The Next Ore Team considered that the only method to move the rock might be to drill and blast, but this would have presented a problem. The narrow ridge would allow for very little burden to contain a blast, and the main substation for the entire mine was in the 6190 Complex just below the 6190 knob. Damaging that substation with flyrock could shut down the entire operation.

The team continued experimenting with different techniques and found that a single shank ripper from a dozer could be used as a chisel at the end of the excavator arm. The excavator could then use the shank to chip chunks off the crest of the knob and break open cracks that would naturally develop in the brittle rock. Although slow going, this method was used until the knob and ridge were mined down to the point that a drill was brought in and the rock was safely blasted.

Work continued on the 6190 knob well into October. For the mine employees, it was one more change to the skyline that people had to get used to because it definitely changed the face of the operation. Removing the 6190 knob was an achievement in itself that took a significant amount of resilience and innovation to actually complete the work safely and with no loss of equipment. Figure 4.22 shows the work on the 6190 knob after enough rock had been removed to create a working area for the drill.

Rebuilding the 10% Haul Road

July was filled with an upsurge of projects for the Next Ore Team. The remediation of the head scarp was making good progress with the advent of production blasting, and the Bingham Shop demolition was near completion. Work had commenced on the intermediate scarp, the 6190 scarp, and the E5 bench cleaning, each of which could be considered a major project in its own right. Although these projects were just getting started and the team was still going through a learning process, senior leadership and the Next Ore Team could see tangible results. The frustrations of June gave way to a sense of excitement after seeing that all the preparation was starting to pay off.

One of the next major projects to get started was the rebuilding of the 10% haul road. Work on the 10% haul road was broken down into three tasks. The first priority was to cut a slot for the haul road where the Bingham Shop had once stood. Work on this task could commence before the head scarp was stabilized, and cutting in the slot significantly reduced the amount of fill required to rebuild the road across the slide area. The second task was working on the fill crossing the Manefay void. This work could not commence until the head scarp was stabilized. The last step for reestablishing the 10% haul road was to mine the debris that was covering the 10% on the other side of the void.

The 10% Slot

The team could not start filling the void left by the Manefay because of the No Go Zone. However, the most recent design in July called for a slot to be cut into the scarp where the Bingham Shop had been located, which would become the beginning of the new 10% ramp. The Next Ore Team put together the MOC and obtained Mine Technical

Figure 4.23 • Starting Construction of the 10% Slot

Review Team and MSHA approval to start construction of the 10% slot. The MSHA approval for a modification to the 103(k) Order was obtained on July 10 and work began on the 13th.

The same rules applied to the 10% slot as had the mining on the head scarp and 6190 knob: people were not allowed in the No Go Zone, with the exception of blasters with special geotechnical monitoring and procedures. The plans called for remote-controlled or autonomous drills to drill the blastholes inside the No Go Zone, while the holes outside the No Go Zone would be drilled with traditional manned drilling methods. From there, remote- controlled dozers would push the crest into the void left by the Manefay. An existing Hitachi 5500 hydraulic shovel would then be used to mine the slot down to the final grade for the new 10% ramp by having the dozer push material to it while sitting outside the No Go Zone. Figure 4.23 shows the remote-controlled drill and dozer operating in the 10% slot.

The Next Ore Team was completing challenging projects in a timely manner, many types of which had never before been attempted at the mine, and they were becoming proficient at recognizing and writing risk assessments for new situations. The one situation that was not recognized in advance was the blasting conditions for constructing the 10% slot. Blasting was required to construct the slot because of the hardness of the rock. The 10% slot blasting required extra precautions because some of the blasts were within 330 feet of the Operations Building in the 6190 Complex. In addition to the location, the chance of flyrock was increased given that ramp construction requires short holes that are easier to overload because the amount of stemming is reduced. *Stemming* is the dirt or gravel that is put on top of the explosive in the hole to confine the energy and prevent it from going up in the air.

The first few blasts went off without incident, but the blast on August 9 sent a large amount of flyrock more than 1,100 feet toward the buildings at the 6190 Complex. Five buildings were hit with flyrock. The largest piece weighed 75 pounds and went through the women's shower and continued on into the men's bathroom located in the Operations Building. In addition, 10 pickup trucks were damaged. More than $100,000 in damage was realized from this one blasting incident.

Although the blast caused property damage, no people were at risk because of the required 1,500-foot evacuation radius around blasts, which is the standard practice for all blasts at Bingham Canyon Mine. Consequently, the entire 6190 Complex had been evacuated for the blast. However, serious events, such as blasting incidents, result in full investigations to determine the cause and prevent further occurrences.

The investigation into this incident identified that 8 to 10 of the 205 blastholes had been overloaded and were not designed correctly, creating the flyrock. Working further to determine the root causes, the incident had more to do with inexperience and not recognizing the critical nature of the blast. First, the blaster that loaded the holes did not have experience in measuring the shallow drill holes of small diameter that were sensitive to the high rates that the holes were loaded with explosives. In addition, the blasting engineer who designed the blast was filling in for the vacationing senior blasting engineer and was not experienced with the design program. Finally, even though the blast was close to buildings, there was not a procedure in place to double-check the pattern and loading with a senior blaster.

As a result of the blasting incident, the blasting procedures were changed, especially for critical blasts around facilities. The company made the decision to do more cross training so blasting engineers would have the appropriate training to fill in when needed. Ultimately, this proved to be very beneficial given that the senior engineer retired approximately one year later and the younger engineer had developed into a capable blasting engineer by that time.

Figure 4.24 • Start of Fill Bridge 2

Fill Bridge 2

July and the first three weeks of August proved to be an especially effective time for the Next Ore Team. By the end of this period, the team had reached its goal of finishing the E5 safety benches, the 10% slot had been constructed, and the team was ready to start filling in the rest of the 10% haul road. Progress was being made on both the intermediate scarp and the 6190 knob.

On August 5, MSHA approved a 103(k) Order modification so that work could begin on the first section of Fill Bridge 2 (FB2) in the bottom of the pit. FB2 was a ramp that started at Code 30 and followed the west wall of the pit and would eventually connect to the E4 North ore that would be uncovered later in the year. Figure 4.24 shows the beginning of FB2 on the back side of the pit bottom.

Fill Bridge 2 was critical to the remediation effort. The ore in the north side of the E4 cut was left without road access because the ore benches in E4 South, which normally would have been used to reach the area, had been mined as part of the ore recovery. FB2 had to be constructed so that when the ore on the north side of E4 was uncovered, the ore could be hauled to the in-pit crusher.

The plan for the August 5th 103(k) Order modification allowed work in the No Go Zone at the bottom of the pit by building up FB2 so that the elevation of the working area would be above the run-out if the head scarp failed. The extent of the construction of FB2 toward the north was limited so that if there was a failure and the debris came flying off the highwall like it did during the Manefay, it could not reach the people and equipment working on the top of FB2.

Unlike the rest of the remediation work, FB2 was constructed by members of the First Ore Team who were working in the bottom of the pit as part of their normal mining process. The First Ore Team had to mine some of the Manefay debris at the bottom of the pit to uncover additional ore in E4 South. By being able to start constructing FB 2, the team was able to put the debris material to a beneficial use instead of just stockpiling the material and rehandling it again at a later date.

Moving to Manned Operations

During July and early August, as the remediation work was progressing at an increasingly faster pace, there was significant progress being made on the geotechnical analysis of the stability of the head scarp. This analysis was critically important to determine how much the Next Ore Team had to remove from the head scarp to make it stable and safe. If too little was removed, then people could be put at risk. If too much, then the remediation would take longer than necessary and the mine could have a gap in ore production in early 2014.

The geotechnical analysis work had to go back to the very basics of evaluating the stability of the head scarp, given that the failure of the Manefay demonstrated that previous analysis was not sufficient. All aspects of the analysis were challenged and reevaluated, and a back analysis of the Manefay failure was performed. This back analysis could then be used as a way to calibrate the inputs going into the head scarp stability. The analytical work was done in two phases. The first phase was a back analysis that had to show that the models accurately represented the perceived situation with the input data and resulting movement as well as the factor of safety of the head scarp. When that phase was completed, the models could be used to do forward analysis to calculate the factor of safety for various unweighting plans for the head scarp.

Although the firm that had performed the geotechnical analysis before the Manefay failed was doing the analytical work for the head scarp stability analysis, a whole new set of geotechnical analytical tools were used compared to the earlier work. Instead of just using two-dimensional (2D) Flac software and limit equilibrium modeling programs to evaluate the stability, the firm also employed Flac3D and Clara 3D (and later SVSlope) to evaluate the head scarp. The use of 3D analyses was a strong recommendation of the Mine Technical Review Team (MTRT) and supported by the internal geotechnical experts.

The consulting firm ran a variety of scenarios for the head scarp stability with different strengths on the Manefay fault, rock mass strength, and quartzite joint strength, as well as the percentage of rock bridging that would be realized. Each of these variables, along with the presence of ubiquitous joints (release joints that go throughout the rock mass), were the key physical properties that went into the geotechnical models.

The initial results from Flac3D showed that using the values that went into the Manefay back analysis did not reflect reality for the head scarp. The mass would have significantly more movement than what was being measured with the monitoring systems, so the head scarp required a higher strength in the model to reasonably calibrate.

Ultimately, the MTRT and Geotechnical Team settled on a case that most closely reflected the actual movement on the head scarp. The strengths from this case were then used for the geotechnical analysis to determine the factor of safety using the 3D modeling techniques.

The results of the Clara 3D model for the head scarp delivered a factor of safety that was marginal for the mining that was completed in late June. The first question that the MTRT and Geotechnical Team needed to resolve was: What short-term factor of safety was required to start manned operations on the head scarp? The second issue was: Should the factor of safety for the long term be the same or greater than the short-term factor of safety?

After multiple discussions and debates, the MTRT supported design criteria that met industry standards for 2D analysis and were slightly higher than the factor of safety used for the designs before that point in time.

It should be noted that having a stable Flac3D model was the first criterion that had to be met for mining the head scarp. Flac3D is a numerical modeling program and takes into account more failure mechanisms than the 3D limit equilibrium programs.

The geotechnical team and MTRT also believed it prudent to increase the long-term criteria for the head scarp. This was justified by two aspects. First, work on the head scarp would have the greatest amount of monitoring possible, so the Geotechnical Team would be able to detect any instabilities and alert the operations in time to protect people. In the long term, there would be a good monitoring program in use, but not as intense as during the remediation work. Second, the head scarp was on the Manefay bed—the cause of the largest mining landslide in history. Consequently, having a higher factor of safety seemed prudent.

Through the first half of July, the Next Ore Team was building designs for the remediation work on the head scarp to reach the point that manned equipment could be used as soon as possible. Josh Davis, the engineer assigned to do the remediation planning, was doing the designs and working directly with the consulting geotechnical engineers to calculate the factor of safety for each bench to determine the ultimate amount of material that had to be removed from the head scarp. One of the interesting outcomes of this work was that the resulting factor of safety was dependent on which calculation method was used in the SVSlope program. The two calculation methods that were compared were Bishop's and Spencer's. The factor of safety results for mining down to the 7190 elevation using Bishop's method was slightly lower than the factor of safety from the Spencer method. The MTRT agreed with the Geotechnical Team's recommendation to use the more conservative Bishop method for calculating the factor of safety.

After the Manefay failed, 18 designs were made for the head scarp remediation. By July 19, Josh Davis and the consultants had determined that the head scarp had to be mined down to the 7190 elevation to reach the desired factor of safety and the 7140 level to reach the long-term factor of safety.

The Next Ore Team now had definitive targets for the head scarp to work toward—7190 for allowing manned equipment below the head scarp and 7140 as the ultimate elevation. Knowing the goal motivated the Next Ore Team to get the work done as quickly as possible. In particular, it was important to get to the point where manned equipment could be used. The team had learned a lot about remote-control operations, but work would be much more efficient by having operators on the equipment.

However, to go to manned equipment before the head scarp reached the long-term factor of safety, the MTRT required that the Next Ore Team put an intensive monitoring program in place. The Geotechnical Team planned for not only the type of monitoring systems that would be used, it had also required procedures for actions to be taken if some of the monitoring systems were not operational. For instance, the head scarp had both an IBIS radar and a GroundProbe radar system. If one radar was not operational because of maintenance or other issues, then work could continue but repairs were expedited. If both of the radars were not operational, work was stopped and the area would be evacuated.

The Next Ore Team also had to do time studies of how long it would take operators to evacuate an area when they were told to do so. Typically, this was in the 5-to-10-minute range. On occasion, the team would perform evacuation drills to ensure that people knew the evacuation routes and how to react if there was a problem.

The weakest link in the monitoring systems was the data network. If the data network went down, then multiple monitoring systems would be affected and the bottom of the mine evacuated. The monitors themselves might remain operational, but if the information could not get to dispatch or the geotechnical engineers, the entire system was considered not to be operational. The Next Ore Team had multiple occasions when it evacuated its work area because the data network went down. A majority of the time the system went down during inclement weather, but each time, the affected area was quickly and efficiently evacuated.

The Next Ore Team reached the 7190 elevation on the head scarp just before August 19. On the night of the 20th and morning of the 21st, the MTRT met with Megan Gaida, who had recently been promoted to superintendent of the Geotechnical Department, and the Next Ore Team to review the progress of the unweighting as well as the monitoring and evacuation plans. They agreed with the Geotechnical Department's assessment (with a few minor caveats) that the No Go Zone could be eliminated and manned operations could commence.

On the afternoon of August 21, the principal advisor for safety and health met with MSHA, and the 103(k) Order was modified so that manned operation could begin below the head scarp. In the case of both MSHA and the MTRT, there had been significant communications before August 21, so both groups were expecting that the Next Ore Team was about to reach the targeted goal. Therefore, they were ready to sign off when the target was achieved.

August 21 was a key date in Bingham Canyon Mine's road to recovery. By allowing manned operations on and below the head scarp, work could commence on dumping in the rest of the 10% haul road and, more importantly, uncovering the ore in the E4 North sector. Since April, ore production was limited because of the three shovels that had been destroyed. The 10% haul road was important because it allowed more shovels to access the pit and would permit the transport of overburden above the in-pit crusher and conveyor out of the mine. The overburden mining would hopefully prevent an ore gap in 2015 and beyond. But opening the 10% haul road was not the bottleneck to continuing ore operations. Stripping the E4 North was the critical piece to making sure there was not a gap in the ore being delivered to the downstream operations. If that area could not be uncovered by the end of 2013, the mine would run out of ore in the first part of 2014.

The race was on, but at least now the Next Ore Team had a chance to meet its goal of uncovering the Next Ore by Thanksgiving.

Reestablishing the 10% Haul Road

With the 103(k) Order modification permitting manned equipment to be operated below the head scarp, the priorities for the Next Ore Team shifted to completing Fill Bridge 2, uncovering the E4 North ore, and reestablishing the 10% haul road. Work would continue on the scarp remediation but with a lower sense of urgency because it was no longer a critical path item.

Reestablishing the 10% haul road meant the mine could start to transport the debris and waste out of the pit. Haul trucks and support equipment could also be maintained at the new Copperton Shop, and the time to get people and supplies into the pit would be significantly reduced. The First Ore Team in the bottom of the pit had done a tremendous job of using the Keystone access to keep the mine running and nearly reaching the budget ore tonnage rates, but having the 10% haul road open again would be more cost-effective and much easier to manage.

The 10% Haul Road Fill

Work on the 10% haul road fill started on August 22 after the plan was approved by the MTRT and MSHA the day before. Figure 4.25 shows the 10% haul road fill being built with manned equipment and the 10% slot leading down to the fill area. The MTRT and Geotechnical Team had three concerns when building the 10% haul road fill:

1. The original design included areas where the fill could potentially impound water, which could contribute to instability of the 10% haul road fill.
2. Parts of the 10% haul road fill could be constructed on remnants of the Manefay bed material that could make the road unstable.
3. The rate of advance when building the 10% haul road fill could be too fast, which would make the fill unstable.

Figure 4.25 • Start of the 10% Haul Road Fill

Having a stable 10% haul road was critical to the Bingham Canyon Mine. The road would be used for the entire life of the mining operations and needed to be reliable. The planned life was until at least 2030, so the 10% haul road fill had to be constructed for longevity.

It was not known whether water would be impounded by the low areas in the 10% haul road fill or if it would just soak into the Manefay debris material. The concern was that impounding water could result in increased water pressure that could contribute to failures in the 10% haul road fill. To address the impounding water issue, the 10% haul road was redesigned to fill in any low-level areas so that any runoff water could be managed.

Filling in these low spots had the added benefit of creating a relatively flat area along the 10% haul road that could be used to park equipment and store road-building materials, as well as provide a place to drill geotechnical and dewatering holes. Having a flat piece of ground in a large open pit mine is a luxury that the operations can benefit from. The downside of the additional fill material is that it added weight to the 10% haul road fill, which could contribute to instability. Additional geotechnical analysis was completed on the extra fill material, and the resulting factor of safety for the 10% haul road fill met the new geotechnical design criteria.

The 10% haul road fill was built through the void left by the Manefay failure. The failure occurred on the Manefay bed, and in some areas, pieces of the bed remained in place. In other areas, the bed was totally eroded away. The problem was that it was difficult to see where the original Manefay bed remained intact because it could be covered by debris material.

The MTRT took a field trip to look for Manefay bed material in the footprint of the 10% haul road fill. During the outing, clear evidence of the Manefay bed was observed. Consequently, the Next Ore Team devised a plan to drill and sample the footprint of the 10% haul road fill. The arrow in Figure 4.26 shows a drill rig drilling the 10% footprint ahead of the fill operations. A number of areas in the footprint were identified as containing Manefay bed material. The Next Ore Team brought in dozers and excavators to remove the material before it was covered by the 10% haul road fill.

Figure 4.26 • Drilling Manefay Material in 10% Footprint

Rate of advance refers to how many feet per day the 10% haul road fill could advance as part of its construction. If the road advanced too quickly, the fill material would not have time to compact and would increase the likelihood of a slope failure, which can be extremely dangerous to people and equipment working on the fill. A low rate of advance would result in extending the time to complete the project as well as creating a potential gap in ore production in the future.

The rate of advance is highly dependent on the height of the fill being placed, the foundation conditions, degrees of confinement, the method of construction, and the material used for construction. The greater the height, the slower the maximum rate of advance. The Next Ore Team had approximately 1,500 linear feet of road to build with fill material to reach across the Manefay failure area. At the original estimated advance rate of 20 feet per day, getting across the Manefay would take 70 days. If everything went according to plan, the Next Ore Team would not reach the other side until November 1, and then they would still have to remove Manefay debris material to complete the 10% haul road.

The MTRT expressed some concern about the 20-foot-per-day advance rate, especially when the Next Ore Team reached the areas of highest fill, which were approximately 225 feet on the down dip side. The Next Ore Team and Geotechnical group would have to monitor the stability of the 10% haul road fill closely and adjust their rate based on observation instead of a fixed distance per day.

Based on those concerns, the Next Ore Team found ways to maintain the high advance rate and ensure that the 10% haul road fill was stable and built to high standards. They came up with a plan, the first part of which was to bring in a full-time geologist, Rudy Ganske, to work with them to construct the 10% haul road fill. Rudy's job was to go throughout the mine and find the best material for building the 10% haul road fill. The material could come from stockpiles, intact rock that could be mined, or any of the Cornerstone faces that were being mined. Some of the areas were a long distance from the 10% haul road fill, but it was better to have a bit higher cost up front for extra haulage rather than long-term problems with the road.

Figure 4.27 • Building the 10% Haul Road Fill

The second part of the plan was to have the best operating practices possible for building the 10% haul road fill. The Next Ore Team operators and Rudy and came up with a process that the fill would be built with a combination of coarse and fine materials so they could get the best compaction. Every truck load that was dumped on the fill would be directed by a dozer operator who had experience in building roads and ramps.

Rudy scoured the mine and found an excellent blend of blocky and fine material that would not break down over time. Rudy also inspected the footprint of the fill areas to make sure that the Manefay bed had been removed and there would be a good foundation at the base of the fill. Figure 4.27 shows the dozer operator directing a truck driver where to place his load of dirt, and the multiple colors show the variety of rock types used to build the ramp.

The Next Ore Team followed the advice of the MTRT by closely monitoring the stability of the 10% haul road fill and adjusted their advance rates based on how the fill was performing. The fill was doing remarkably well, and the advance rates were increased because of that performance.

Testing the 10% Haul Road Fill

As the Next Ore Team continued to increase the advance rate, Zip Zavodni expressed concern that the advance rate might be too much. He was skeptical that the fill was performing that well. On Friday, the 13th of September, an intense and violent rainstorm hit Bingham Canyon Mine. The work on the fill for the 10% was only half complete, which made it one of the most susceptible times for damage if there were any problems or irregularities in its construction.

The day it rained presented an opportunity to test the stability of the 10% haul road fill in extremely wet conditions. The sudden storm came over the Bingham Canyon Mine with extremely heavy rain and lightning from the southwest side of the pit and was heading toward the 6190 Complex. The storm hit the 10% haul road fill with its full force at a time when the construction was at the point of highest fill, and thus the most vulnerable to damage or failure.

At the time the storm arrived, two other events were going on. First, a meeting was in session with most of the mine's managers, superintendents, and several supervisors. They were discussing plans for a visit the following Monday from Rio Tinto's Executive Committee and Board of Directors. The group was coming to see the Manefay damage and witness the recovery efforts firsthand. It was a rare event for the mine to have such an important group visit, and the meeting was held to make sure everything was ready.

The second event was that the blasters were getting ready to shoot a blast on the 6190 knob just above the Operations Building where the meeting was being held. The blast was not scheduled to go off until later that afternoon, so the blastholes were being loaded but were not ready to be set off.

Because the lightning storm was going to reach an area where there were loaded holes, the blaster followed set procedures and evacuated everyone within 1,500 feet of the blast area. A radio call was made to the blasting superintendent that the 6190 Complex had to be evacuated and the announcement was made in the meeting. The entire Mine Management Team picked up their belongings and headed for vehicles. There was no discussion or debate, just a calm but hurried response.

The rain had already started before the meeting broke up, and as everyone left the building, they were instantly drenched. The rain was coming down so fast that a sheet flow of water carried rock debris across the access road leading down to the Lark Gate—something that had not been witnessed by anyone currently at the mine. Everyone from the 6190 Complex safely evacuated the mine and traveled to the mine entrance. By the time the team had made it to the Lark security gate less than two miles away, there was no rain and the road was dry. The storm was small but amazingly intense.

The weather stations at the mine recorded between 2.6 and 3.6 inches of rainfall in the mine in less than 3 hours. This little storm qualified as a 1,000-year rain event in that part of Utah, and there was damage throughout the area. Interestingly enough, the areas east of the mine received little to no rain that day, but were hard hit the next day when there was flooding near the Rio Tinto Regional Center.

Figure 4.28 • Helicopter Landing

At about the same time that the 6190 Complex was evacuated, a decision was made to stop the ore mining operations in the lower pit as a precaution in case the heavy rain would destabilize the head scarp or damage sections of the Keystone access road. The pit dispatchers and supervisors notified everyone working in the lower pit to rendezvous at the muster point near the in-pit crusher. The rain was coming down in a torrent and the Keystone access road—the only road in and out of the pit—acted like a channel for the water to flow into the open pit. The Keystone access road became impassable with deep mud and multiple washouts, preventing the Ore Team from totally evacuating the mine.

The rain stopped after a few hours, but the people remained at the in-pit crusher until the Keystone access could be repaired. Although the weather had cleared and the operators could have returned to ore operations, a decision was made to keep everyone at the muster point instead of returning to work to limit the possibility of having someone getting hurt when emergency evacuation would be difficult. The crew remained at the muster point until nearly the next shift change.

Suspending ore production in these conditions was an extra precaution since there was a secondary evacuation method for removing injured employees from the mine once the rain stopped. Plans had already been made for helicopter transport from the bottom of the pit in case of emergency and a helicopter test landing had been conducted a few months earlier. In July, the mine had contacted AirMed, an air medical transport service of the University of Utah Health Care, to arrange a test to see if their helicopter could land on a haul road in the bottom of the pit if there were an emergency to evacuate an injured employee. A larger landing location in the pit had been covered by the Manefay debris. With the extra time that it took to exit the pit via the Keystone access road (and the possibility that the road could become blocked), having this secondary egress was critical. On July 3, a landing pad was identified on the haul road and the emergency transport helicopter was given landing instructions. Kennecott's Emergency Response Team members as well as members of the Greater Salt Lake City Unified Fire Authority were on hand when the AirMed helicopter came in to land. The test landing and subsequent take-off proceeded as planned, and everyone at the mine was more comfortable with the knowledge that if someone was hurt, they could be safely evacuated. Figure 4.28 shows the arrival of the AirMed helicopter.

The Keystone access road repairs were not completed until almost the end of the day shift on September 13th. The day had been one of the most unusual during the recovery time frame after the Manefay, with a 1,000-year rain event, evacuation of the 6190 Complex, employees trapped in the bottom of the pit, and lost production. But with all that had occurred, there were no incidents or injuries, the fill for the new 10% haul road withstood a major challenge, and everyone returned home that evening. All in all, it was a very successful day.

For all the damage and problems caused by this small but intense storm, the 10% haul road fill basically received no damage. Water was rushing down the slopes, but when it got to the area that could potentially impound the water, the debris and material on top of the Manefay bed was so porous that it allowed the water to soak in and drain below the 10% haul road fill.

The day after the storm, Zip Zavodni came to the mine to do a thorough inspection of the 10% haul road fill as well as examine some slips in the Manefay failure area. He was expecting to find evidence of cracking and at least some instability in the fill of the haul road. Based on his report, what he found instead was

> ... the 10% ramp fill was in excellent shape. The material blend/quality appears to be near optimum to maximize consolidation and limit ponding. The practical lack of settlement cracks on a recently built thick fill zone with some heights approaching 225 feet is most unusual.

Based on Zip's inspection after the rain, having discovered no stability or erosion problems, the Next Ore Team continued to increase the advance rate, especially as the fill started to reduce in height. The high level of monitoring was maintained as well as control of the material going into the fill.

By September 26, the fill portion of the 10% haul road was nearly complete, and it was ready to tie in to native ground. This was an important milestone that was completed nearly a month sooner than projected. There was still a lot of work to do, given that the area past the tie-in point was covered with nearly 100 feet of Manefay debris. But the team was one step closer to getting equipment into the bottom of the pit and returning to normal operations.

Completing the 10% Haul Road

Once the fill was complete, clearing the debris on the 10% haul road could begin. This debris was not only from the original Manefay slide, it also included debris material from cleaning the safety benches above the 10% haul road. In total there were nearly two million tons of debris that would have to be loaded and hauled away with trucks and shovels.

The tie-in of the 10% haul road fill represented a short-term opportunity to move a Hitachi 5500 hydraulic shovel to the bottom of the pit before the newly commissioned giant P&H 4100XPC AC (no. 99) shovel started mining the fill material. A narrow road had been built over the debris material, and a relatively small and diesel-powered shovel such as the no. 63 shovel, a Hitachi 5500, could traverse the road and get into the pit bottom. However, if it was not done quickly, the 99 shovel would start mining and remove the access.

Moving the 63 shovel to the bottom of the pit accomplished two things. First, it could increase ore production—the highest priority after keeping people safe. The existing ore-producing 68 and 64 shovels could actually maximize the production to the crusher and concentrator when both shovels were operational. But every time one of the two shovels went down or had to do other work, ore production was affected. The addition of the 63 shovel meant that it could back up the two shovels and increase ore production.

Second, the 63 shovel could help remove the Manefay debris from the 10% haul road or pit bottom when it was not needed for ore removal. This would accelerate how quickly the new P&H 4100 shovel could get in the bottom of the pit and open up more ore to maintain production. Although the mine could reach planned ore production rates most of the time without the shovel, it would increase the reliability of ore production and enabled the removal of waste from the pit bottom so overburden mining could resume.

Debris mined by the 63 and 64 shovels in the pit bottom was used to construct Fill Bridge 2. This fill bridge was required to be in place before mining of the next ore could commence and therefore prevented a gap in production. Having the 63 shovel in the bottom of the pit was very beneficial to returning the mine to near-normal operations, and the mine was able to nearly meet the budgeted ore production in October of 2013.

The new 99 shovel was constructed in record-breaking time and commissioned on August 24 so it could start mining fill to build the 10% haul road and then remove debris from the road. As can be expected, the shovel experienced some electrical and mechanical problems during the shakedown period in the first two to three weeks of operation. The maintenance crews and vendors worked tirelessly during this time to get the shovel to stable operating conditions so it could uncover the 10% haul road. Figure 4.29 shows the 99 shovel mining Manefay debris from the 10% haul road with the support of dozers and excavators pushing material and cleaning the final highwall.

On October 27, the 10% haul road was completed. The road was finished sooner and better than anyone could have expected. Figure 4.30 shows the final 10% haul road. The road—even with the fill over the Manefay failure area—was stable and in excellent condition. The mine could have continued mining ore for several more months without the haul road, but getting the 10% back in place helped tremendously to get the mine back to normal operation. No longer would the employees and suppliers have to use the Keystone access road to get in and out of the pit.

Figure 4.29 • 99 Shovel Mining the 10% Haul Road

Figure 4.30 • Completed 10% Haul Road

Large equipment could go in and out of the mine as needed. Haul trucks and support equipment could be repaired in the Copperton Shop instead of on the gravel pad at the bottom of the pit. Life was much closer to normal for the men and women of Bingham Canyon Mine—but more hard work remained.

Uncovering the Next Ore

The final steps for the Next Ore mandate was to uncover the ore buried by the Manefay debris on the north side of the pit in the E4 cut and complete the construction of Fill Bridge 2 so the ore could be hauled to the in-pit crusher. Much of this work was undertaken by the First Ore Team headed by Nate Foster because there would be some overlap of the equipment required for ore mining and remediation work. Figure 4.1 shows the relationship of the Manefay debris on top of the E4 ore that had to be removed to start mining the encumbered ore. The figure also shows FB2 that had to be constructed to allow mining of the ore once the north side of E4 was uncovered.

A limited amount of work on FB2 was started in early August but was discontinued by the middle of August when the construction reached the point where the operation could be affected by a failure of the head scarp. After the head scarp was stabilized on August 21st, manned equipment could be used below it. Uncovering the E4 cut with dozers and excavators as well as building FB2 could proceed in earnest. Work began almost immediately in clearing the benches below the 10% haul road. Just as with the E5 benches above the 10%, dozers and excavators would clean the bench and cast or push the material to the bench below. The difference now was that all of the work could be done with manned operations.

Approximately 10 million tons of debris material would have to be pushed to the bottom of the pit to uncover the next ore. At the beginning of the process, the benches were narrow and the progress was slow because of the limited working room. But with each successive bench, more and more working room was created because the debris would extend further and further out into the pit. As a result, most of the material would have to be rehandled multiple times because on each bench there was material from the previous benches. Figure 4.31 shows dozers pushing debris material into the pit as they worked down to the E4 North ore.

The deeper the debris cleaning went, the more the team learned and the more equipment could fit into the operation. Excavators would be used to remove debris material up against the highwall, at which point the dozers would then push the debris down the slope. With each successive bench, the dozers would have to push more material for a longer distance. Consequently, more and more dozers would be brought in to move the debris material. Figure 4.32 shows how the excavators cleaned along the highwall as the dozers pushed the debris down into the pit.

In September, the debris material being pushed by the dozers started to reach all the way to the bottom of the pit. This meant that the dozers were pushing out on a nearly 700-foot slope. As the team was able to add more dozers to push the Manefay debris, production rates increased, and once again, the rate of advance became a concern, just like it had with the 10% haul road fill. The issue for uncovering the E4 was that the team could not pick and choose what type of material would go into the fill; they had to use the Manefay debris material. By mid-September, the team began slot dozing with the dozer fleet, and the advance rates increased even more. By late September, the rates had increased to cleaning more than 25 vertical feet of highwall in a week, although cracks appeared in the dump at times, which indicated instability.

Figure 4.31 • Cleaning Benches to Uncover the Next Ore

Figure 4.32 • Excavators and Dozers Working to Uncover the Ore

Figure 4.33 • Location of Buttress

Nate and his team put together an E4 Fill Settlement Plan, which was part of the Trigger Action and Response Plan, or TARP, to manage the advancing E4 debris removal. The TARP was similar to the Manefay Response Plan that was developed before the Manefay failure and provided instruction for operations as conditions changed. The Geotechnical Team performed inspections at least once a day, and the operators were trained to inspect for instabilities in the fill material. These included the obvious cracking and also a not-so-obvious bulging of the slope. The dozer operators were also trained to measure the slope angles on the fill to make sure the face did not become over-steepened.

The TARP called for moving dozers out of areas that showed stability problems. The working areas were broken down to sections, and the dozers would switch sections if needed. Based on recommendations from the MTRT, the TARP also called for allowing time for the slopes to rest and stabilize. The plan dictated that after the dozers pushed for three days on any one section, there had to be at least six hours of idle time to monitor the slope. The team had to schedule all of this work to keep the dozers working effectively between the various sections.

The geotechnical engineers began monitoring the fill slope with radar systems. Radars are not normally used for fill slopes because of the amount of movement in the fill operating process. In this case, the geotechnical engineers could monitor the wall because of the area's size, and after some instability was detected, the dozer equipment was moved to another area while the unstable area was monitored. The movement in the wall decelerated as the slope settled in and became more stable. Based on the intense monitoring program, the fill slopes stabilized within six hours, so the procedure became that the dozers could not move back to an area that had showed instability for at least six hours after the radars detected movement. By being able to move back and forth across the working area, there were few times that the dozer push work was totally shut down.

The TARPs and procedures went through a thorough review in the Management of Change process and received support from the MTRT. The Operations Team was very proficient with its monitoring and management of the high-fill slopes. Each week, the high-risk slopes became shorter and shorter as the team worked down the highwall, which helped reduce the risk of a failure, yet more and more material was being moved. To further stabilize the slope, the Mine Management Team decided to build a toe buttress with waste debris. When they were not mining ore, debris was mined by the two hydraulic shovels, 63 and 64, and hauled to a ramp as close as possible to the fill slope. The new ramp provided a buttress on the toe of the fill (Figure 4.33), which provided additional stability and lowered the risk. This was debris material that had to be moved to uncover ore in E4 South to maintain ore production, so it was work that needed to be done, and the buttress provided a short haul distance.

The team continued to improve their methods, but by the end of September they were still only dropping the debris material by 25 feet per week because of the increasing rehandle. They needed to find another way to increase the rate if they were going to prevent a gap in ore production. To that point the team had been working only during daylight hours because of the concerns around the fill stability. The one potential solution was to be able to work 24 hours a day instead of just daylight hours. The team started exploring options by putting as many lights on a dozer as possible. They ended up installing four lights above the blade and another three on each side of the dozer above the cab. The team then had to get approval from the Geotechnical group, Management group, and the MTRT to switch to nighttime operations.

The first response from almost everyone was—NO WAY. It was just too important to be able to see the cracks. However, after much discussion and debate, it was decided to at least investigate to see how much and what could be seen with the improved lighting. Josh Reese, one of the geotechnical engineers, was nominated to ride with the dozer operator after dark and evaluate how well the operator could see any cracks developing in the fill material.

Figure 4.34 • Dozer Lighting

What Josh found surprised many people. The new lights did a very effective job of lighting up the work area. In fact, in some ways, the cracks were even easier to detect because they created shadows, which provided more contrast than the cracks seen during daylight hours. Figure 4.34 shows the increased lighting on the dozers. The Next Ore Team made some adjustments to the lights on the dozers and relocated light plants on the crest of the slope to better delineate bulging in the fill face, but the reality was that working at night would not significantly reduce detection of instabilities in the fill material. Armed with this new information, the Next Ore Team was given approval to start nighttime operations, and the debris started to drop by 49 feet per week. The rate was much better, but the team needed even more to meet the Thanksgiving goal.

As October progressed, more and more dozers and excavators became available as other critical jobs, such as the scarp remediation and 10% haul road, were completed. As the equipment became available, they were deployed to the E4 fill, and the fleet in that area was increased from approximately 10 dozers and excavators to more than 30. It was an impressive sight to see 26 dozers pushing in slots beside each other while four excavators cleaned the highwall. Stability of the fill was becoming less of a problem because the slope had shrunk to fewer than 400 feet and the toe buttress was in place. Figure 4.35 shows 20 dozers and excavators moving the Manefay debris that covered the E4 North ore near the end of October.

By November 1, the debris material was dropping 125 feet per week and the team was within 100 feet of the top of the of the next ore block—E4 North. Over the first two weeks in November, the team dropped the debris the final 100 feet and pushed in the remainder of Fill Bridge 2 with the Manefay debris. Now the final connection was made to the E4 ore block. Figure 4.36 shows the Next Ore Team tying in Fill Bridge 2 to the top of the north end of the E4 ore.

Figure 4.35 • Dozing Manefay Debris

On November 13, the first load of the next ore was mined in E4 North and transported to the in-pit crusher. This was the milestone that everyone had been striving for—uncovering the ore so the mine and the rest of Kennecott's operations would not experience a significant gap in production. It was an astonishing accomplishment to not only reach a goal that almost everyone thought was impossible, but to deliver the results nearly two weeks earlier than projected! Figure 4.37 shows several members of the Next Ore Team in front of the first truck load of ore mined in E4 North. Figure 4.38 is the first load of next ore being dumped.

Figure 4.36 • Uncovering the Next Ore

Figure 4.37 • First Truck Load of Next Ore

Figure 4.38 • First Load of Next Ore Being Dumped

In the seven months between the Manefay failure and the first load of ore produced by the Next Ore Team, the men and women of Bingham Canyon Mine undeniably *rose to the occasion.* They faced and safely met overwhelming challenges in record-breaking time. They found ways to stabilize dangerous scarps, remotely demolish a building, build a new haul road, and remove debris to uncover ore. In every case, the hard work, dedication, and innovation of the Kennecott employees and their vendors resulted in viable solutions. And those solutions kept people safe while being productive. In the time after the Manefay, the people of Kennecott worked more than a million person-hours without a reportable injury or lost-time accident.

The accomplishments after the Manefay were a result of the work of an entire company and not just one group. Many groups rose to the occasion: independent reviewers (such as the MTRT), senior leadership within Kennecott and Rio Tinto, and a wide variety of technical experts as well as community and government organizations. Without the help and support from these groups, and many more, the timing and results would have been much different. It required an organization willing to listen to employees, as well as the government officials and technical experts inside and outside of Rio Tinto who acted decisively and always kept safety as their first concern. The success also required the vision and leadership of senior management at Kennecott and Rio Tinto. If they had not given the teams the leeway to make decisions and instilled confidence that they would do the right work, the goals would not have been met. In short, every person in every part of the organization understood the goals and objectives, which started with keeping people safe, maximizing production, and preventing an ore gap, and they did everything in their power to meet those targets.

Lessons Learned from Uncovering the Next Ore

Chose the right leader. Choosing the right leader made a huge difference for the Next Ore Team. Providing that leader with mentoring and resources support is critical to his or her success. After the leader is chosen, bestow your trust in him or her to accomplish the job, even during rough times.

Create a thinking culture. The leadership and processes used during the Manefay remediation created a thinking culture where employees from all levels and parts of the company were actively involved in solving problems and reaching a goal. There are several components to creating that culture, but some of the ways that it was accomplished at Bingham Canyon Mine was to create difficult goals, ask for everyone's help to meet the goals, provide the time and expectation that the problems will be solved, and give people recognition when they do solve the problems and accomplish the goals.

Carefully analyze options. At first glance, blast casting seemed to be the fastest way to remediate the head scarp. But drilling the block would be extremely time consuming, requiring different equipment and a significant amount of traditional mining. Most importantly, it could have created additional risks to people and the operation, as well as a significant delay in production if it did not work as planned. Making the right decision to mine the head scarp in a traditional method was critical to the success and speed of the remediation work.

Do not let smaller projects become the bottleneck. Even though the Next Ore Team had key milestones that they had to complete, such as remediating the head scarp, they continued to find resources to finish smaller projects, such as the Bingham Shop demolition. Staying on top of the smaller projects prevented them from becoming major bottlenecks at a later time.

Step back to do proper risk assessments and plans. When faced with a new situation, step back and do a proper risk assessment and develop a plan, as the Next Ore Team did when the dozer fell over the crest of the head scarp. The result of a well-planned response to an emergency situation is that it builds greater confidence and trust among team members.

Double-check designs of less-experienced engineers. When senior personnel are not available for critical designs, such as blasting of the 10% slot, ensure that a system of checks and balances are put in place to check critical designs.

Establish a second method of egress. Even though there was only one access road in and out of the pit after the Manefay failure, the team found a second method of egress in case of an emergency. The team ensured that an air evacuation helicopter could land in the bottom of the pit. Although this method of egress was not utilized, the knowledge and understanding that it could be used ensured that the team was ready if it was required. It also helped ease some of the worries and concerns of those working in the pit who were concerned because there was initially only one route of access.

Review options fully. When the option of changing the lighting for the dozers was brought up so work on the debris material above the ore could continue 24 hours a day, it was almost universally dismissed. However, after actually putting the lights on a dozer and evaluating what the dozer operators could see as far as viewing cracks and the working area, it was apparent that the work could proceed safely. There are two take-a-ways from this—first, to fully evaluate your options, and second, to be persistent when you know your alternative has merit.

Fully use the tools. The Next Ore Team derived significant benefit from tools such as Lean Boards and the Management of Change process. The reason that the Next Ore Team was successful is because their leadership saw that the tools could help ensure that all the right steps were done in a very uncertain situation. In that way, the process was seen as a tool and not just a program that had to be done. The success of many programs hinges more on the belief and attitudes of the people implementing the program than the program itself.

Preventing Another Manefay

Chapter 5

One of the questions that can be asked after a large event such as the Manefay is whether the organization has learned lessons to prevent a similar event in the future. For Kennecott Utah Copper, the opportunity to answer this question came all too soon. In July of 2013, it was discovered that a second large mass might be at risk of failing once the Manefay had been remediated—and this failure could be even more devastating than the Manefay. Not only did the company have to feverishly work to keep from running out of ore by the end of the year, they had to start the unweighting of the second mass.

While tackling this new challenge, lessons continued to be revealed. The first lesson was to apply previous knowledge to identify a new potential risk, and understand the long-term options and costs to address an identified risk. When mitigating risks, changes in the expected conditions should be thoroughly evaluated. Established criteria should be followed unless reevaluated before operating beyond the set criteria. Management should strive to maintain consistency of institutional knowledge of pertinent events to new management.

Identifying the Fortuna Risk

On July 22, 2013, Zip Zavodni and Martyn Robotham, two of the most experienced and capable geotechnical engineers within Rio Tinto, and probably the mining industry, were on an inspection tour of the Bingham Canyon Mine. Since the landslide in April, the geotechnical focus had been on the Manefay, and they thought it was time to take a closer look at the rest of the pit. During the inspection, they noticed an extensometer that was being used to measure changes in a crack on the surface of the mine south of the Manefay and asked the geotechnical engineer who was giving the tour to take a closer look.

What they found was a large crack that was several inches wide and parallel to the highwall. The crack appeared to be wider than the normal deformation cracks that are typically observed in a large open pit mine. Deformation cracks often appear after large-scale mining operations as the rocks and soils relax and try to fill in the void of the open pit.

The extensometer had been installed in 2012 when the crack was first discovered by the mine's Geotechnical Team. When discussing the crack with the mine's geotechnical engineer, it was obvious that there had been enough concern to install the extensometer. A few deep holes were drilled and time domain reflectometers (TDRs) were also installed to monitor movement below the surface of the ground. The engineer reported that at least one of the TDR cables had been sheared off underground at a bed called the Fortuna and had stopped working, which indicated some subsurface movement. However, the surface movement at the face of the highwall was within expected movement rates. As a result, the geotechnical team was not overly concerned about the crack and TDRs, so they continued to monitor the area. In fact, the movement around this area had slowed significantly after the Manefay, primarily because there was much less mining in the bottom of the pit.

Before the Manefay failure, the Fortuna was seen as a potential problem, but on a much smaller scale than the Manefay, so the Fortuna was already being closely monitored. All of this work on the Fortuna monitoring had been done before the Manefay failure and had not been reevaluated in consideration of what had happened with the Manefay. There had been failures on the Fortuna in the past, and there was concern regarding a relatively large 1.7-million-ton failure based on the pre-Manefay geotechnical analysis. The mine's geotechnical engineers had not considered a massive failure like the Manefay, especially since the movement was almost zero at the time—but Zip and Martyn had more reservations. The Fortuna bed had very similar characteristics to the Manefay in that it had a significant amount of "gouge" material above and below it. This provided an area of clay that was weak in which the rock above the Fortuna could slide.

After the inspection, Zip and Martyn traveled back to the Rio Tinto Regional Center (RTRC) offices in South Jordan where Zip's office was located. I had just left my office in the RTRC and was walking down the hallway as they approached. Zip and Martyn stopped me and asked, "Do you know about the Fortuna?" I replied, a bit puzzled, "The Fortuna?" I had come to Kennecott only six weeks before the Manefay failure and was totally consumed trying to prepare for and recover from that event since arriving at the mine. Although I had heard of the Fortuna, I did not totally understand all of its characteristics.

Zip and Martyn disclosed what they had seen and expressed their concerns. They talked about the crack and hypothesized that it might be close to the outcrop of the Fortuna, which could indicate that a large mass was moving on top of the Fortuna—just like what had happened to the Manefay. They also relayed the facts about the TDRs being sheared in the area of the Fortuna bed and how the mine's geotechnical group did not seem overly concerned. Then they warned me that if the Fortuna was moving and it failed, it could be of the same magnitude as the Manefay! In addition, the Fortuna was located above and intersected the 10% haul road, in-pit crusher, conveyor, and tunnel. If a massive failure were to occur on the Fortuna, all ore operations would be halted for a number of months, not weeks like the Manefay. This would have an even greater impact on the mine and could potentially result in its permanent shutdown. Figure 5.1 shows the location of the Fortuna bed outcrop (purple) and the Fortuna dike (blue) in the Bingham Canyon Mine. The crack is at the top of the pit, similar to the Manefay.

In the weeks after the Manefay, there was one message that was perfectly clear from both Kennecott and Rio Tinto senior leadership: another Manefay-size failure was not acceptable. In fact, Rob Atkinson, the chief operating officer for the Rio Tinto Copper group in London, had been at the mine less than two weeks earlier and had made a point of saying just that.

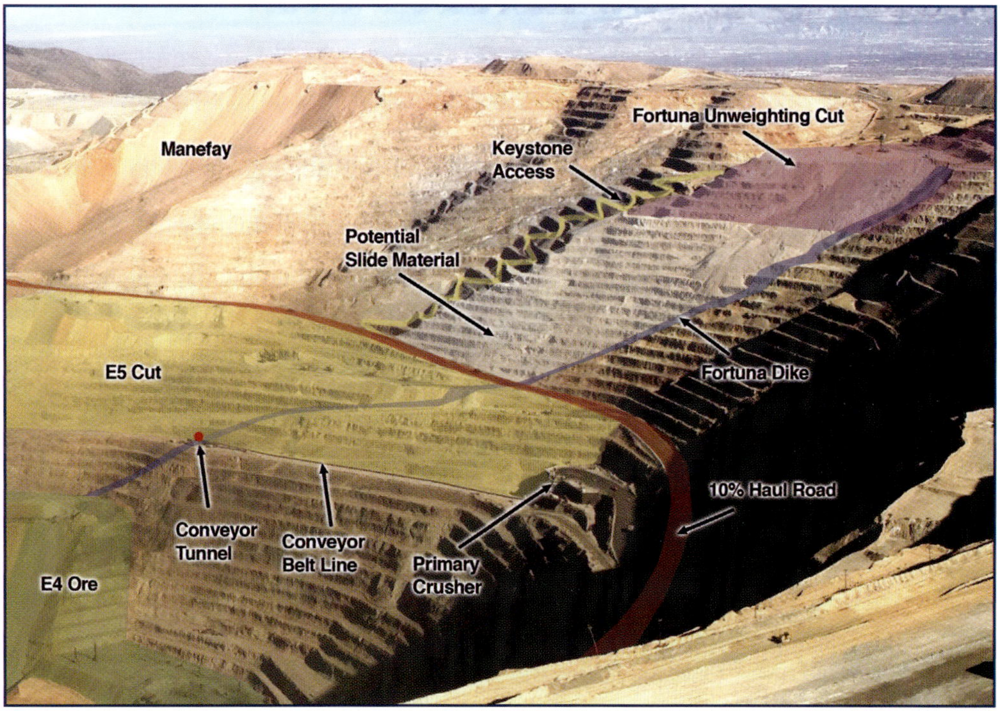

Figure 5.1 • Fortuna Location and Infrastructure

After listening to Zip and Martyn describe the evidence and their concerns, the next step seemed fairly clear—we had to know whether or not the crack was on the Fortuna outcrop. If it was not on the Fortuna, there may not be a problem, but if it was....

After leaving Zip and Martyn, I went directly to the mine planning superintendents, Jon Warner and Joan Danninger, who then brought in two mining engineers, Jon Heiner and Ed Woods. Their task was straightforward—locate the crack in relationship to the Fortuna bed, and if the crack was in the same location, they were to build a plan to unload as much of the Fortuna as possible. At that point we did not know much about the potential failure mechanism or whether unloading would be necessary, but having a plan just in case would be better than not having one.

A short time later, the engineers returned with the locality of the crack. The crack was positioned exactly at the outcrop of the Fortuna. That location and the fact that the TDRs had been sheared indicated that the entire rock mass could be moving as a unit. This information, coupled with the relatively small movements that were observed on the Manefay before it failed, indicated that the Fortuna could become a big problem. The mining engineers immediately started working on a plan to unload the Fortuna.

No one likes to be the bearer of bad news, especially with limited data or if the problem could be particularly large. It can be more comfortable to wait until you have more information and greater confidence that there really is a problem and the scale is understood. After all, if you say there is a problem when there really is not, you risk "calling wolf," resulting in lost credibility and creating undue anxiety. However, if it is a problem and you delay notification, then the opportunity for solutions can be lost. That afternoon, the potential issue of the Fortuna was reported to Matt Lengerich, the mine's general manager, who then reported to Stephane Leblanc, who had recently been promoted to managing director for Kennecott and was taking over many of Kelly Sanders' duties.

Jon Heiner and Ed Woods developed wrote the unweighting plan for the Fortuna, and on August 1 it was presented to the group of mine planning engineers, the Geotechnical Team, and Zip Zavodni. In the design, approximately 44 million tons of waste could be unloaded from the Fortuna. Although the design told us how much material could be removed, at that point there was no geotechnical analysis to tell how much *should* be removed to make the Fortuna stable. We could only guess how much to unweight the Fortuna since the results could range anywhere from zero to more than 44 million tons to make the Fortuna stable. Unfortunately, the geotechnical engineers and consultants were still focused on the Manefay, and it would be October or November before the Fortuna analysis could be completed.

The geotechnical engineers from the mine were still not overly concerned about the Fortuna and recommended waiting until the analysis was complete before starting the unweighting. Unweighting is expensive, and to remove the 44 million tons would cost tens of millions of dollars—a lot of money if it was not needed. Zip and the Technical Services Planning group were more in favor of starting the unweighting as soon as possible.

The Plan Going Forward

That evening, I met with Matt Lengerich to discuss options for the Fortuna. My recommendation was to start unweighting as soon as possible. We would not have the answers for more than two months, so we would be risking several million dollars. However, by starting now, we had a much better chance of preventing a failure if it was a problem. There was no movement on the highwall at that point in time, but as soon as the 10% haul road was completed, we would start mining waste material that would be located on the toe of the Fortuna. This would be the time of highest risk for a failure; so basically, we had a limited time to do the preventive work. The other option was to wait until the analysis was complete. If the analysis showed we had a problem, we would have to delay mining the waste at the toe of the Fortuna, which could result in a gap in ore production at a later date.

At first glance, this would have seemed a difficult decision—to spend several million dollars that might not be required in hopes of preventing a future problem or to wait to get more information. In reality, the downside was too great. A large failure would have been catastrophic, because it could take out the 10% haul road, main crusher, conveyor system, and tunnel. Such devastation could be the end of the mine and the loss of value would be measured in billions of dollars. If delaying the start of unweighting resulted in a gap of ore production for a few additional months, the cost would be measured in hundreds of millions of dollars. Matt's decision was to start the unweighting as soon as possible. As time would show, this turned out to be a great decision.

The Mine Planning Team presented the plans for the Fortuna unweighting to the Mine Management Team on August 2nd. Several projects had to be completed before mining could commence, such as moving a secondary rock crusher and preparing the area so the shovel would have a working space. Power lines had to be moved and cables run so the shovel would have electricity. On August 17, the shovel was in place and unweighting of the Fortuna began. Figure 5.2 shows one of the early plans to unweight the Fortuna. The mine designers found a way to unweight even more material from the Fortuna. The red and green areas represent a 60-million-ton unweighting cut. The zigzag line below and to the left of the shaded areas is the Keystone access road that was critical to ore production after the Manefay.

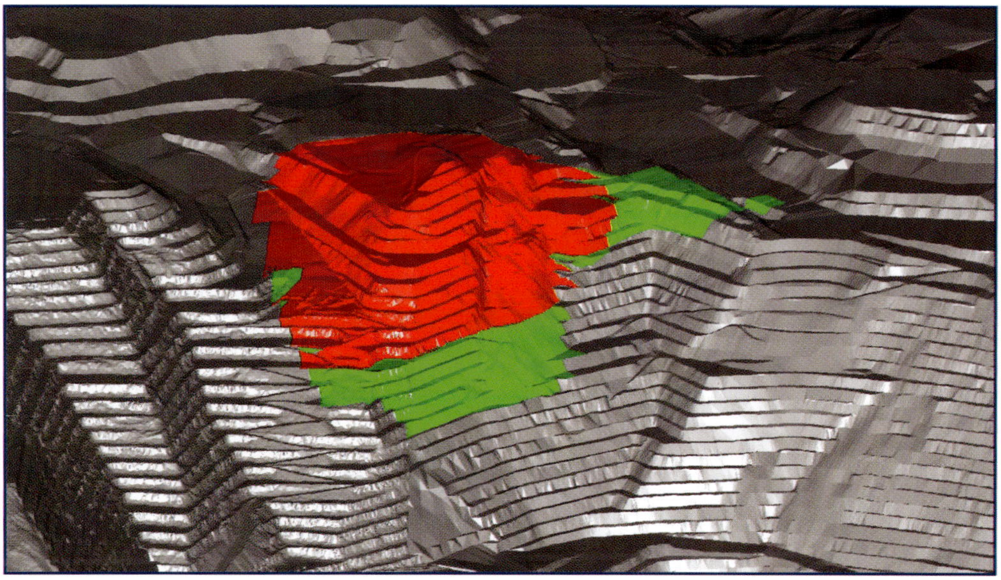

Figure 5.2 • Early Fortuna Plan

By mid-August, the structure of the Technical Services Team at the Bingham Canyon Mine was changed. The Geotechnical and Geology Departments were combined with the Mine Planning Department into a new Technical Services Team under my management. Megan Gaida was promoted to geotechnical superintendent and Jon Warner was promoted from his temporary position to full-time superintendent for short- and medium-term mine planning. Joan Danninger continued to be in charge of the Long-Term Planning group.

By having all of the Technical Services group report to one manager, it was easier for the various specialties to work together as well as break down silos. This proved to be beneficial because coordination between the group members would be important as the analysis work proceeded.

As manager of the new Technical Services Team, one of my first priorities was to formalize the Mine Technical Review Team (MTRT). Rohan McGowan-Jackson from the Senior Leadership Team had been advocating the formalization of an independent team to review the technical work, and this was an opportunity to make the change. The group had been working somewhat informally without a written charter or mandate. Mark Button and Martyn Robotham from the Rio Tinto Technology and Innovation (T&I) group and I put together a formal charter for the MTRT. The charter encompassed an expanded membership that included a hydrologist and a mining engineer. Up to that point, the MTRT was strictly focused on the Manefay geotechnical issue. By formalizing the group, we were able to expand the team's purview to any technical concern facing the mine. This was important as we started to work on the Fortuna, because the MTRT's input and support would be needed on a regular basis for many areas other than the Manefay remediation.

By the beginning of October, the first round of geotechnical analyses was being completed on the Fortuna by a geotechnical consultant who analyzed the Fortuna using three-dimensional (3D) modeling with both the SVSlope and Flac3D modeling programs. Based on the movement that had occurred in the past, it was assumed that the mass was at equilibrium, meaning that it was at risk of failure if conditions changed. Equilibrium indicated that the factor of safety was 1.0 or slightly higher and therefore well below the criteria that had been used by the MTRT for mining of the Manefay head scarp, so remediation plans would be needed.

In the analysis, the geotechnical consultant assumed that the Fortuna bed was almost dry, but this was based on very limited data. The rock strength was reduced to minus one standard deviation below the mean of the drill-hole data sample rock mechanic tests, which was low, but comparable to the Manefay back analysis. The team also modeled the Fortuna as a stand-alone bed as far as the potential failure mechanism. Although these assumptions seemed reasonable, over time they would have to be reevaluated based on actual results of the slope stability.

After the geotechnical analysis indicated that unweighting was required, the next step was to determine how much material would have to be removed. Ed Woods, one of the senior mining engineers who had developed the initial unweighting plan, traveled to Tucson, Arizona, to work directly with the geotechnical consultant. Together they evaluated a number of unweighting scenarios to determine how much material would need to be removed.

Ed and the geotechnical consultant engineers evaluated six unweighting options by calculating the factor of safety at different bench elevations. The goal was to reach a factor of safety similar to the Manefay head scarp for the ultimate highwall on the Fortuna. Just like the Manefay head scarp work, a somewhat lower factor of safety was required during the unweighting process compared to the long-term requirement. These targets were reviewed and supported by the MTRT based on the back analysis regarding the stability of the Fortuna bed.

Ultimately the analysis showed that for the long term, 37 million tons of material would have to be removed to prevent the potential failure of more than 68 million tons if the Fortuna failed. However, a significant amount of the Fortuna would have to be removed before mining could commence at its toe when the 10% haul road was rebuilt and the waste could be hauled out of the mine to the mine's ultimate dumps.

Fortunately, the decision had been made to start unweighting the Fortuna in August. In the 2½ months from the start of mining to the end of October, nearly 10 million tons of material had been removed, which had increased the factor of safety enough to meet the short-term criteria. This meant that mining of the E5 waste on the toe of the Fortuna could start as soon as the shovels could be moved into the pit without additional unweighting delays.

The 10% haul road would be completed by the end of October, and there would be a drive to start mining the overburden at the toe of the Fortuna on the E5 as soon as possible to uncover the next cut of ore. But there was a problem—mining of the E5 would lower the factor of safety. So the question was, could the unweighting of the Fortuna stay ahead of the mining of E5?

The Race

The E4 ore mining was progressing, but if additional ore was not uncovered, the mine would deplete all uncovered ore in 2015, even if the Next Ore was uncovered by the end of 2013. To prevent depletion of the ore, the next cut of ore would have to be uncovered below the E5. To accomplish this, the E5 overburden would have to be removed so the conveyor belt could be moved and the ore below the E5 could be mined. The E5 was at the toe of the Fortuna, and work could only proceed as fast as the Fortuna was being unweighted. If the E5 was worked too fast, the Fortuna could fail. If the E5 was worked too slow, there could be an ore gap to the downstream operations.

At the beginning of October, Kennecott's senior leadership made the decision to start a mandate team to mine the E5 waste. Elaina Ware would manage the team. Mining had been delayed for seven months on the E5 and could not be resumed until the 10% haul road was completed so the shovels could be returned to the pit and waste could be trucked out of the pit.

In conjunction with mining the E5 waste, the plans before the Manefay called for moving the in-pit conveyor in November and December of 2013, which was based on mining down to the level that the conveyor was located. The Fortuna was exposed in the benches between the 10% haul road and the conveyor, and therefore these areas would be impacted by the movements of the Fortuna. If a massive failure occurred on the Fortuna, the 10% haul road, in-pit crusher, conveyor, and the portal to the tunnel that went through the mountain would be destroyed.

The E5 Mandate Team was modeled after the success of the First Ore and Next Ore Teams as a self-contained group that included a manager, top supervisors, and the best operators, as well as a full-time mining engineer for technical support. The team was assigned the best equipment available to help ensure there were no bottlenecks for achieving their goals. Other than mining ore, the E5 Mandate work became the highest priority for the company.

As with much of the work on the Manefay, Stephane Leblanc set an aggressive goal for the E5 Mandate Team—to have the conveyor moved by June 1, 2014. This target required the conveyor to be moved two months earlier than first anticipated in the plans after the Manefay. Somehow the E5 Mandate Team would have to cut two months out of the schedule between November and June by either accelerating the mining of the overburden for the conveyor or relocating the conveyor at a faster rate, or a combination of both.

The mine's Technical Services Team kept the MTRT up to date on the progress of the geotechnical analysis. After the mining of the E5 started, the factor of safety for the stability of the wall would start to drop, unless the unweighting of the overburden on top of the Fortuna could progress fast enough to maintain a stable wall. The MTRT strongly supported the policy of not letting the factor of safety drop below the criteria that had been set. The MTRT communicated directly to the managing director to ensure that everyone in the company understood the importance of not mining the E5 too soon.

Unweighting would have to progress at a very high pace to keep up with the extremely high rates that were planned for mining the E5 waste. It would be a race between the Fortuna unweighting and E5 overburden removal—if the Fortuna Unweighting Team lost, then neither group could win. To keep ahead of the E5 schedule, the Fortuna Team had to mine approximately 13 million tons of material from November 2013 through January 2014. In the meantime, the E5 Mandate Team was scheduled to mine approximately 6.9 million tons in the same amount of time, but the team was also gearing up to achieve much higher rates to cut the time to uncover the conveyor bench to achieve the goals that had been set.

At the end of October, the Next Ore Team work was starting to wind down. The 10% haul road had been reestablished, the head scarp and intermediate scarps were completed, and the safety benches above the 10% were cleaned. Work was progressing smoothly and ahead of schedule on cleaning the debris off the E4 ore that was covered up on the north side of the pit and that work would be finished by bottom-of-pit operations before the Thanksgiving goal. At the beginning of November, Matt Lengerich made the decision to move Cody Sutherlin and Don Mallet from the Next Ore Team to manage the Cornerstone overburden mining and Fortuna unweighting. Cody, Don, and the entire Fortuna production team understood the importance of staying ahead of the E5 Mandate Team's schedule—they did not want to be the bottleneck that slowed down production. The challenge with the Fortuna was that it involved a small area with short faces. Operationally it was difficult because the team would have to start a new bench at least once a month. Starting a new bench was much slower than normal operations, so it negatively impacted production rates. Figure 5.3 shows the close proximity of the shovels when mining the Fortuna.

Figure 5.3 • Fortuna Unweighting

In the end, the race did not turn out to be much of a contest. The Fortuna Team exceeded its scheduled production by mining more than 12.7 million tons between November and the end of December—nearly a month ahead of schedule. The team had averaged over 210,000 tons a day with two shovels and the support of a front-end loader in a small working area during the months of November and December. This was nearly 25% more than the normal rate for the shovels. Consequently, the Fortuna Team was never a bottleneck to the E5 Mandate Team in achieving its goals.

By the end of March 2014, the Fortuna Team completed the remaining 13 million tons of unweighting. With the entire 37 million tons of material mined, it was believed that the ultimate factor of safety had been achieved and that any potential problems with the Fortuna had also been averted. Nevertheless, the movement of the Fortuna was closely monitored by the geotechnical engineers—just in case. The shovels and equipment were moved out of the Fortuna cut and returned to Cornerstone to continue uncovering the next pushback, and everyone believed we would be returning to a steady-state operation.

E5 Operations Continue

At the end of 2013 and into early 2014, the E5 Mandate Team was on track to meet the post-Manefay schedule to uncover the conveyor bench, but it had not made much progress in accelerating the rates to catch up the additional two months. The team was using three shovels, but the bottleneck was the intersection from the 10% haul road and the width of the cut. At the intersection with the 10% haul road, the haul trucks would have to make the turn going across traffic and the turn was greater than 180 degrees. This turn slowed the flow of traffic and led to congestion that in turn resulted in additional decreased traffic flow. Figure 5.4 shows the paths that the haul trucks had to travel around the intersection going to the E5 operations.

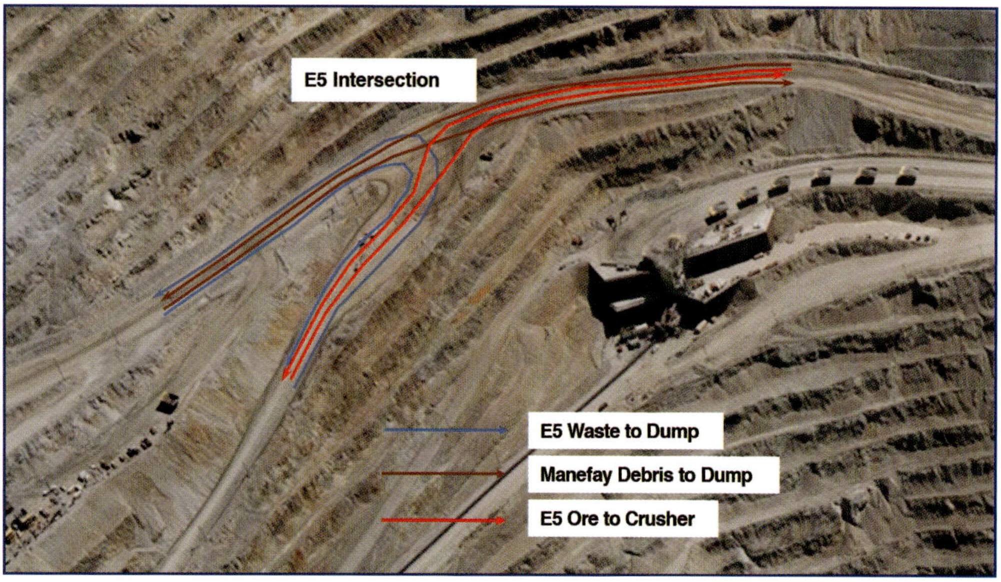

Figure 5.4 • E5 Intersection Congestion

In addition to the troublesome intersection, the E5 bench was very narrow at approximately 225 feet wide. This width provided much less operating room than other parts of the pit, such as the E4 South cut where the First Ore Team had a bench that was closer to 350 feet wide. The First Ore Team was able to reach production rates that were nearly 125% of the initial post-Manefay plan, even when they had multiple shovels in the cut. The narrow bench in the E5 area resulted in conditions where the drilling of blast holes, cleaning of the haul road, and mining with only two shovels contributed to the congestion instead of increasing production.

At times it seemed like the harder the E5 team tried, the more intractable the problem became. Toward the end of December, several of the trucks from the Cornerstone operations were sent to the bottom of the pit in an attempt to boost production. The ore production realized an increase, but the congestion in E5 acted like a regulator and the team did not see a corresponding rise in production. Frustration levels swelled throughout the entire mining team.

In January it was decided that the pit bottom crew, including the E5 and Ore Production Teams, did not have enough haul truck drivers to keep all of their trucks operating. Haul trucks were shut down in the Cornerstone area so that operators could be assigned to trucks in the pit bottom in an attempt to increase production. Once again, adding resources did not solve the problem. The E5 Team was only meeting the original schedule of having the conveyor bench *uncovered* by June instead of the new mandate of having the conveyor *moved* by June.

During this time, the Fortuna seemed relatively stable. The geotechnical team was closely monitoring the movement of the Fortuna highwall. Geotechnical monitoring data have some variability built in because of the sensitivity of the equipment and factors such as changing temperatures. Domenica Cambio from the Rio Tinto T&I group found that by taking averages of prism readings over a week, small trends of 0.01 inch per day or less could be detected and tracked. This was significant because the Manefay was detected at movement of 0.1 inch per day. Therefore, if the mine could start to monitor at movements of 1/10 of the rates that were previously monitored, this would improve predicting and preventing future failures. At a later date, Domenica and the Geotechnical Team started to calculate and report each day a moving average of the movement over a seven-day period. Using this method, trends of a 0.001 of an inch per day were measured and helped in tracking the trend of the movement.

As expected, the movement of the highwall increased when the E5 Team started mining at the toe of the Fortuna. The level of 0.02 inch per day was determined by the Geotechnical Team as the point where the Operations Team would be alerted. This level and process was supported by the MTRT. Figure 5.5 shows the weekly rates for the prisms on the Fortuna mass. If the movement was less than 0.02 inch per day, the data point was in the green zone, which meant that the movement was within expected levels. If the movement went above 0.02 inch per day (the area in red), closer monitoring and analysis were required.

The movement continued to increase until it exceeded 0.02 inch per day, wherein the Geotechnical Team put the Operations Team on notice that if the bed movement was accelerating, mining would have to be curtailed or stopped. By January 10, the movement on the Fortuna decreased and then leveled off, and mining continued without interruption. The 0.02 criterion for increasing the level of concern appeared to be an appropriate measure.

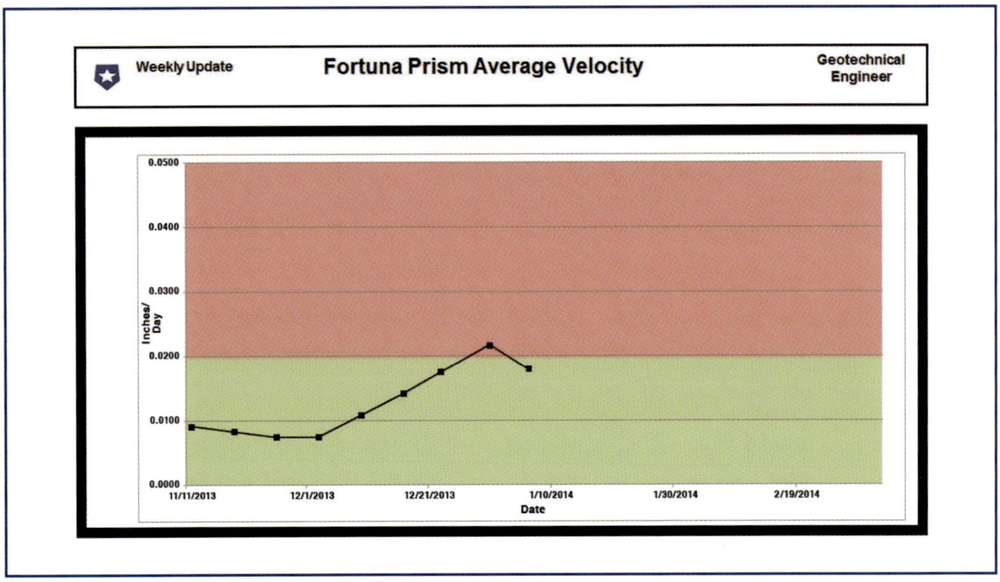

Figure 5.5 • Fortuna Prism Rates Through December 31, 2013

By February there were concerns that the E5 Team would not meet its goal of uncovering the conveyor bench scheduled for April so the conveyor could be moved by June. The E5 Team was focused on eliminating the congestion at the intersection with the 10% haul road and they were starting to see limited progress with a small but insufficient increase in production.

Planning and preparations for moving the conveyor were also progressing. John Thompson had been assigned to manage the actual moving of the conveyor, and his team had developed plans to reduce the time to dismantle and reconstruct it, but he needed a firm date when the bench would be cleared. The Senior Leadership Team affirmed that the date of June 1 was fixed.

Something had to change for the E5 Team to complete the mining of the conveyor bench to meet its goal of getting the conveyor moved by June 1. Although the team did not have to allow a full two months to move the conveyor, at the current rate they would barely beat the June 1 deadline with moving the material off the bench, and this would leave no time to actually move the conveyor. Toward the end of February, a solution was developed by the E5 Team's mining engineers. Instead of mining out the entire bench, the team could focus on only the portion of the benches directly above the conveyor so that mining could progress while the new conveyor was being constructed.

The plan was innovative and provided a pathway to cut 20 days out of uncovering the conveyor bench. The new plan, combined with the improved plan to move the conveyor in less than half the original planned time, meant that the E5 Team could meet its goal. But this also required accelerating the mining of the Fortuna toe at extremely high rates. As Zip had warned earlier, the rapid mining could shock the system and result in accelerated movement. Mining would need to be stopped until the wall stabilized.

The Fortuna was already experiencing some smaller-scale bench failures where mining had recently taken place. This area had historically been a geotechnical problem, and a block just below the 10% haul road had failed, causing a temporary restriction on the road. At that time, the E5 Team had drilled holes in the benches below the block failure and placed railroad tracks and cement in the holes to attempt to stabilize the area. Although this was an area of concern, it was localized, and the overall slope seemed stable at the time.

The E5 Team started the accelerated mining in mid-March. Zip had also warned that the team would need to be very careful when blasting, so many of the same procedures used in the head scarp mining were implemented for the E5 mining. The MTRT visited the site toward the end of March and supported the mining of the E5 based on the reduced movement on the prisms above the 10% haul road. This difference in movement indicated that the movement on the upper part of the Fortuna was not connected to the movement below the 10% haul road, which lowered concern for an overall massive failure.

As the E5 Team started the accelerated mining, the Geotechnical Team closely monitored the highwall movements. Since the end of February, the rates had been above 0.02 inch per day, but each week the values had been up and down and therefore did not show an acceleration trend. However, from mid-March to April 10, there were three weeks of continuous increases and the movement went from less than 0.03 inch per day to more than 0.04 inch per day. The levels were still much lower than the Manefay, but the continuous increase indicated the potential for acceleration—not a good thing. The levels were low, so there was no immediate danger for a large failure, but an action plan was needed. Figure 5.6 shows the acceleration over the three-week period. The Geotechnical Team had added the movements of the other beds near the Fortuna, such as the Winnebago, Congor, and Parnell, and the Parnell was showing acceleration as well.

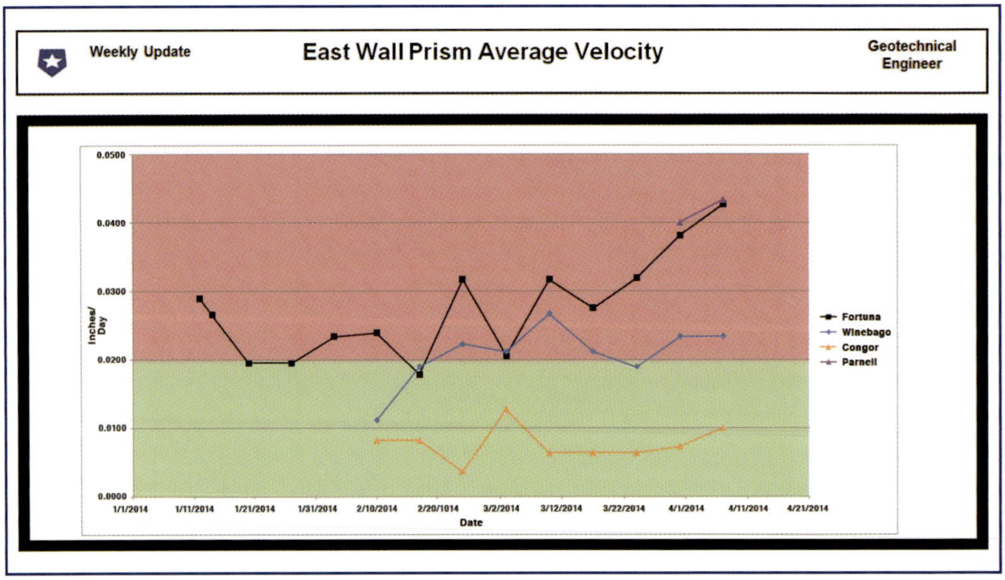

Figure 5.6 • Fortuna Prism Rates Through April 11, 2014

On April 9, mining of the E5 halted. At this point, the Geotechnical Team was in almost constant communication with the MTRT, given that this was new territory for how to handle accelerations on a very small scale. The decision was made to stop mining for three days to see what happened to the movement. If the movement showed deceleration, then mining of the E5 could resume. However, if acceleration continued or increased, mining would not resume until a different plan was developed.

The Fortuna responded to the pause in mining almost immediately. The acceleration stopped and the rates leveled off at approximately 0.045 inch per day. That was almost exactly the response we had hoped for, so the mining of the E5 resumed. Based on the modest acceleration, the geotechnical engineers started to report the daily rolling average movements for the Fortuna instead of once a week. The MTRT recommended placing a threshold of 0.05 inch per day as a limit to mining.

About this time, I approached Matt Lengerich and informed him that there needed to be a change in management. The mine was in the process of performing a technical review for the future of the mine based on the Manefay and geotechnical risks going forward. Over the past four months, I had been managing the technical review and day-to-day technical services work, both of which were critical to the mine. By dividing my time between the two, neither received adequate attention.

My recommendation was that I would dedicate 100% of my time to the technical review and a different manager would be in charge of the Technical Services Team temporarily until a full-time manager could be selected. Matt concurred, and the decision was made. In retrospect, it might have been better if I had retained responsibility of the Technical Services Team at least until the E5 work was done.

After resuming E5 mining on April 12, the E5 Team was driving hard to uncover the conveyor bench by May 1. This would give the team moving the conveyor one month to meet the overall target of June 1. But meeting this goal would not be easy, as geological conditions were still changing. As mining progressed down the highwall, water began to appear in the Fortuna drill holes. The further down the highwall the team worked, the more pronounced the water flow became. In the last few benches, the water was actually flowing out of the hole and over the bench in an artesian flow, indicating that the water was under hydrostatic pressure. The E5 Team was mining the bench so quickly that the water did not have time to naturally drain.

The water affected the operations in two ways. First, it made the operation more difficult by having to work in mud instead of on solid ground. Second, it increased the water pressure on the face of the highwall. The hydrologic studies had indicated there should not be much water in the Fortuna. In fact, when the operations personnel reported the presence of the water, the hydrologist said that was impossible, yet it was definitely present. Water pressure would make the highwall less stable, but the general feeling was that the water must be localized and therefore did not represent a problem.

On April 26, the movement rate of the Fortuna started to increase, but the geotechnical engineers needed at least three or four days of data to determine whether there was a trend for acceleration. By the time the trend could be determined, the movement had passed the 0.05 threshold set by the MTRT. At this time, the E5 team was just one day from completing the mining of the conveyor bench. The Fortuna was not in danger of failing in the short term, and based on the experience where the acceleration stopped as soon as mining stopped (and the fact that meeting the goal was within reach), mining on the E5 continued for the additional day. On May 1, the E5 Mandate Team reached the goal of completing the conveyor bench so work to move the conveyor could start. There would be no mining on the toe of the Fortuna for at least a month, which should give it plenty of time to stabilize.

Although mining stopped on the first of May, the acceleration on the Fortuna did not stop. The movement continued to increase from approximately 0.055 inch per day to more than 0.065 inch per day by May 9. During that time, it seemed as if the collective Technical Services and MTRT groups held their breath. Had the Fortuna been activated to a progressive failure that could not be stopped? Would there be another Manefay? It was impossible to know for sure because no one had ever monitored such large-scale slopes at such small levels before. We were now to the point where changes of a few thousandths of inch in a day seemed to matter. Figure 5.7 shows the Fortuna movement with the three-day suspension and increased rate before the mining was completed for the conveyor move.

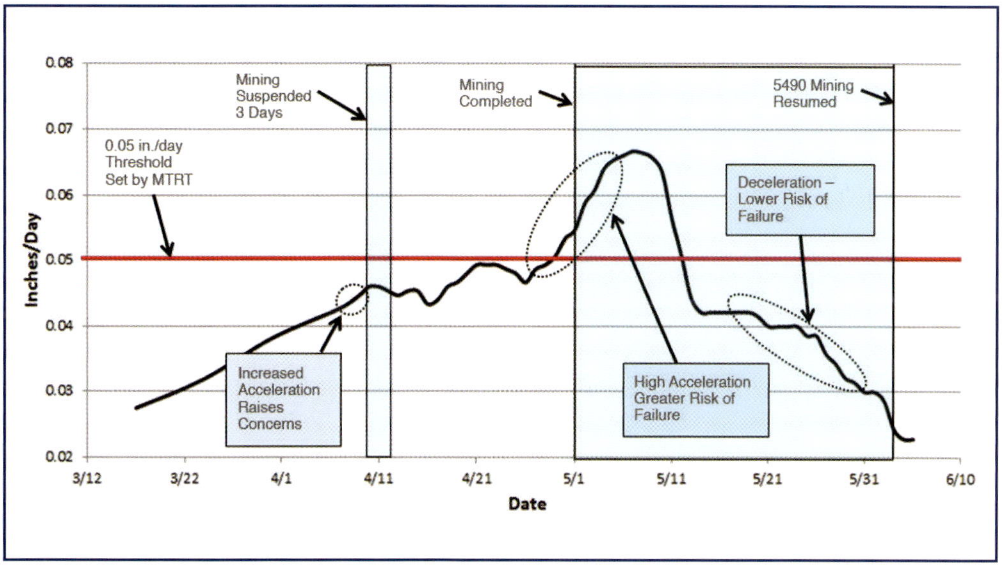

Figure 5.7 • Fortuna Movement in March Through June 2014

On May 10, the acceleration of the Fortuna finally subsided and the daily movement leveled off. When the movement report was released, you could almost hear the collective sigh of relief. With the cessation of mining on the E5, the movement on the Fortuna started to drop. By May 31, the movement was fairly close to 0.02 inch per day. Moving the conveyor was achieved by the June 1 deadline, so the team had met its goal, but not without some anxious moments.

My belief is that there were many ramifications due to the movement of the Fortuna in April and May. In my mind, the first and most important was that if the Manefay had not happened, the Fortuna would not have been unweighted, small movements would not have been measured, and mining would not have been adjusted. Without the Manefay, chances are we would have mined the Fortuna to failure.

The second realization was that the hydrology and geology models were incorrect. Had the mining proceeded at the normal rate, the water could have drained out of the rock, preventing the muddy working conditions, accelerated sinking rate of the bed, and reduced stability of the Fortuna. In addition to the water and sinking rate, there was a corresponding movement of the Parnell bed that was directly below the Fortuna. The Fortuna had not been modeled in conjunction with the Parnell, and further geotechnical analysis showed that a complicated step path failure between the two beds was possible, and that resulted in a lower factor of safety than that for the Fortuna by itself.

The events surrounding the remediation work on the Fortuna confirmed that there were still risks at the Bingham Canyon Mine, but they could be mitigated. Even though the geology of the Cornerstone pushback did not have beds of weak gouge material like the Manefay and Fortuna, the Mine Management Team wanted further analysis to identify possible geological risks.

Detailed geotechnical analyses continued through 2014 and 2015. The Bingham Canyon Mine made a step change in the amount of geotechnical analysis and monitoring that is done to safely operate a large and deep open pit mine. During that time, the Geotechnical Team and consultants working on the Bingham Canyon Mine brought geotechnical analysis and monitoring to a whole new level, with more-detailed 3D analysis, dewatering, and mine planning. The skills, capabilities, and experience of the Technical Services and Operations groups have also dramatically improved to the point that they are on the cutting edge and world class.

Lessons Learned for Preventing Another Manefay

Apply what you have learned. Preventing the Fortuna failure is the best example of applying the wisdom gained from previous events. The Mine Management Team began work on unweighting as soon as evidence of bed movement was observed. The Geotechnical Team started to measure and react to much lower levels of movements than before the Manefay slide, which made a huge difference in preventing a Fortuna failure. The Mine Management Team furthered this process by evaluating the Cornerstone area for geological risks.

Understand the cost of failure. One of the reasons why the unweighting started so early was because the cost of having the Fortuna fail was understood. The general manager risked wasting tens of millions of dollars for the potential benefit of saving hundreds of millions, if not a few billion dollars. If the general manager was only concerned with short-term cost savings, he would have made a different decision, which would have cost the company a tremendous amount of value.

Reevaluate based on actual conditions. When artesian water was observed in the blast holes of the Fortuna, work was not stopped to allow further study and evaluation of the large-scale impact of these changing conditions. Although the operational risks of working in the wet conditions were considered, the geotechnical risks from the unexpected water were not analyzed. The results of the Fortuna movement demonstrate the need to reevaluate changing conditions in a holistic manner.

Consider the sinking rate of the mine. The speed at which the E5 team were mining benches (sinking rate) was incredible, because the mine plan limited the length of the benches and the need to move the conveyor. Multiple benches were removed in less than one month. This sinking rate did not allow for benches to adequately drain, which contributed to the instability of the highwall.

Follow the criteria. Near the end of mining the E5, there was at least one day wherein the 0.05-inch threshold was exceeded and mining continued. Although it is understandable to try to meet a goal that was set and the team was driving to achieve it, there was also a chance that the Fortuna could have gone into a progressive failure that could not have been halted. More discussion and detailed risk assessment should be completed before continuing to operate past set criteria.

Maintain consistency in management and institutional knowledge. There is value in maintaining consistency and institutional knowledge during critical times. A new manager in charge of the geotechnical group did not have the same background and sensitivities to movements in the highwall as the previous manager and therefore would have had a difficult time to see a problem trend.

Why the Manefay Failed

Chapter 6

The question most often asked about the Manefay is "Why did it fail?" Although the answers primarily involve geological forces and how those forces were predicted to behave, there are lessons learned about how known events can be analyzed for additional unknowns and addressed for any company that faces risks. Questions should be focused on how risks could be observed from a new and different perspective, perhaps through an independent reviewer.

The Manefay was a complex active/passive failure mechanism where a mass of material is moving (the active piece) but is resisted by a smaller passive section that is not moving—until the stress becomes so great that the passive section fails (sometimes catastrophically). This type of failure acted significantly different than any of the previous failures at the mine. Because the failure mechanism was not identified, the mine was not designed to prevent the failure during the mining process. For the many decades before the Manefay, the mine had experienced literally hundreds of much smaller and simpler wedge and circular failures. The mine's geotechnical staff and consultants had become very knowledgeable with these types of failures, and their geotechnical analyses evolved to model and anticipate them.

Unfortunately, the active/passive failure was not modeled; therefore, the engineers did not anticipate the failure with their geotechnical analyses. Although the failure was not prevented, the fact that the geotechnical engineers understood that highwall failures were a significant risk enabled them to develop a world-class monitoring system to detect and predict highwall movement. It was these monitoring systems and methods that the geotechnical engineers used to predict the Manefay, thus preventing any deaths or injuries as a result of the gigantic event.

The Manefay could rightfully be categorized as a Black Swan event as described by Nassim Nicholas Taleb (2012) in his book, *The Black Swan: The Impact of the Highly Improbable.* A Black Swan event is a metaphor for a large and unexpected event that is an outlier to what would normally be expected and therefore changes future expectations. Following are the three criteria for a Black Swan event and how the criteria applied to the mine:

1. **The event is a surprise.** This was certainly true for the Bingham Canyon Mine, given that it was a failure mechanism that had never been experienced before and the size was significantly larger than ever realized in any mining operation.

2. **The event has a major impact.** This was true for the Bingham Canyon Mine, as the recovery effort required a major financial commitment. It was also true for all large open pit mines that should now consider the possibility of a large-scale active/passive mechanism failure as part of their geotechnical analyses and highwall designs.

3. **Being the first recorded instance, it is rationalized by hindsight, as if it could have been expected.** Since the Manefay, there have been a multitude of people who have said that the Manefay could have been anticipated using models of non-mining failures. In hindsight that could be true, but based on a long history of highwall failures at the Bingham Canyon Mine, active/passive failures were never an issue, and therefore not considered. In the future, geotechnical and mining engineers should benefit from the knowledge acquired from the Manefay to understand and prevent comparable occurrences.

To understand the Manefay and why it failed, one needs to be familiar with the geologic complexity of the massive Bingham Canyon Mine and the geotechnical analysis that was performed to design the highwalls at the mine. After a brief description of the geology and analysis, the rest of this chapter describes the Manefay failure and the lessons learned in the process of identifying and evaluating a risk.

The Manefay landslide was a complex failure at a mine with intricate geology and geologic structures. This chapter provides an overview to give the reader a general understanding of how and why the Manefay failed and is not intended to be a comprehensive description. Hopefully there will be professional papers or books written to provide in-depth details of the geotechnical nature of the failure that go beyond the size and scope of this book.

Geologic Conditions

The Bingham Canyon Mine has been a major producer of copper, gold, silver, and molybdenum since 1906. As with most large open pit mines with multiple metals, the geology at the Bingham pit is complex and varied. The geology of the mine includes sedimentary rocks that have been folded, faulted, intruded by igneous materials, and subjected to metamorphic alteration (Porter et al. 2012). It is this complexity that resulted in the ore mineralization and the creation of one of the world's most historic ore bodies. This complexity also created the conditions of bedding and folding that when combined with the mining of the open pit and placement of waste piles led to the ultimate failure of the Manefay bed.

One of the most important features of Bingham Canyon Mine is the variety and intensity of folding and faulting that took place 50 to 140 million years ago during the Sevier orogeny (compressive mountain-building event). This large-scale folding and faulting was not only the genesis of the Oquirrh Mountains, it created the pathways that allowed molten rock from deep within the earth to travel close to and at times breach the surface. Around 38 million years ago, the Bingham Stock intruded into the native sedimentary rocks. A *stock* is a large (approximately 40 square miles or less) igneous intrusive that forces its way into the local rock. The Bingham Stock intrusion brought up much of the hot mineralized fluids that would become the Bingham Canyon ore body. Molten rock from the Bingham Stock forced its way further up through the surrounding rock in the form of dikes that cut through the preexisting beds and sills that ran along the beds.

The Bingham Canyon Mine is so large and deep that the various types of geologic structures can be seen in the highwalls of the mine. Figure 6.1 shows a photograph overlaid with the geology of the south wall of the mine. The primary feature is the area shaded in green and red, the Bingham Stock, which is an igneous rock called monzonite because of its higher concentration of feldspars and plagioclase, and relatively lower amounts of quartz. The areas shaded in yellow are quartzite, which was originally sandstone, and has been altered by the heat and pressure of burial and intrusion by the Bingham Stock.

Figure 6.2 shows a view of the northeast area of the mine. This portion is much more geologically complex. The Bingham Stock is still shaded in green and red at the bottom of the photo, but further up is a series of parallel beds called the Parnell, Winnebago, Congor, and Manefay, which are finer grained than the surrounding quartzite. Theses beds have been intruded by molten rock that has formed dikes, such as the Fire King, Starless, and Andy, shown as the tan features cutting through or along the beds (labeled in Figure 6.2). The photo also shows the Fortuna sill, which started as a dike cutting across the beds until the molten rock found a plane of weakness between the Parnell beds and the quartzite above, at which point it started to follow the Parnell bed and created the Fortuna sill.

The stress and strain of the mountain being formed and intrusion of the Bingham Stock forced the ground to move and buckle. What had once been flat beds of sandstone, limestone, and siltstone began to tilt and bend. The heat and pressure from the molten magma that was moving toward the surface changed the sedimentary beds, turning the siltstone beds into hornfels (a tough, fine-grained metamorphic rock) and the limestone into marble.

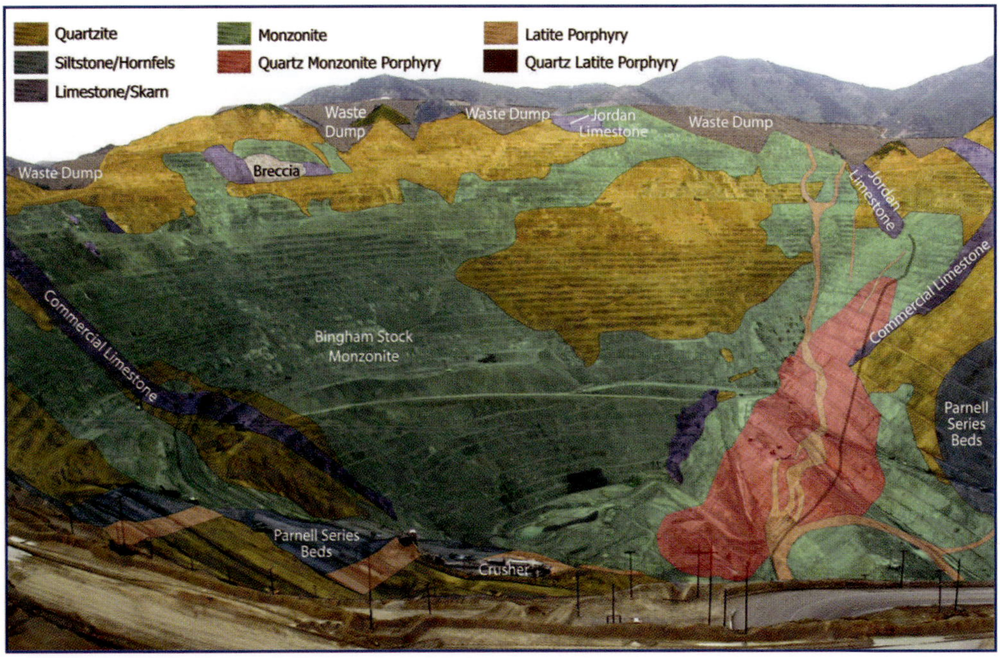

Figure 6.1 • Bingham Canyon Mine Geology Looking South

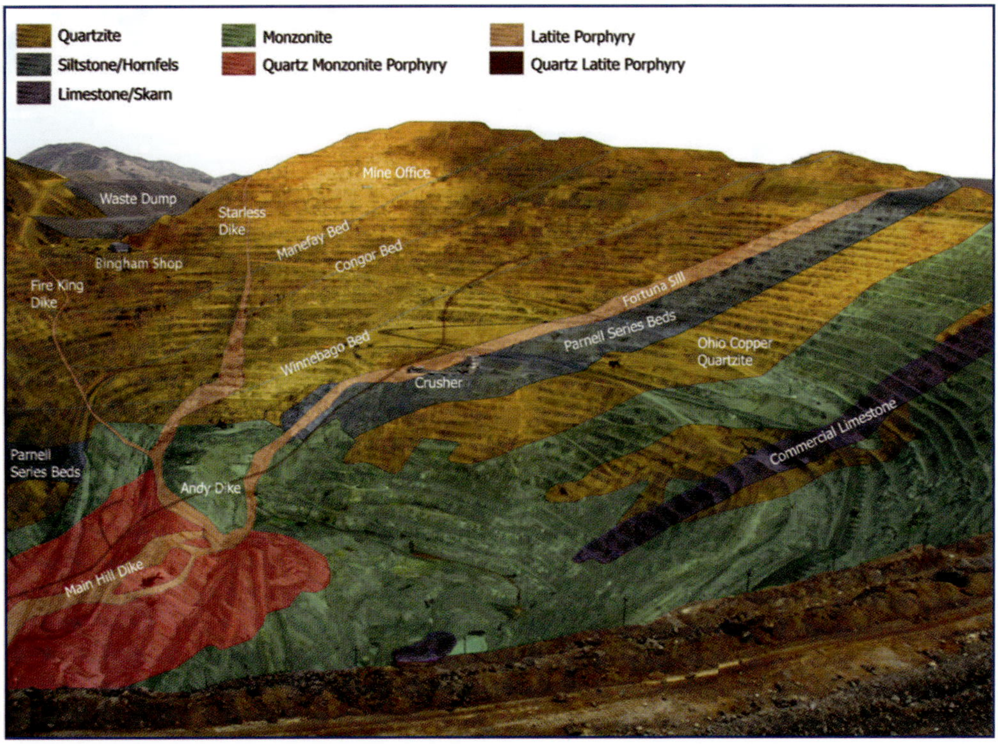

Figure 6.2 • Bingham Canyon Mine Geology Looking North

Contacts between the sandstone and siltstone beds were areas of differential strength (one bed being stronger than the other), so as the earth bent and folded, it was between these beds that experienced the most differential movement, which was critical for two reasons. First, as the movement started between the beds, it reduced the strength between the beds even more. Whatever cohesion the beds had was basically ground away by the tremendous weight and pressure of two rock surfaces moving past each other. In effect, the more movement there was, the easier it became for additional movement.

Second, in addition to the reduced cohesion between the beds, as they moved past each other, they ground the rock contacts to a very fine-sized clay. This clay material, called gouge, is important not only because of its low strength, but also because it has low permeability and therefore reduces the speed that water passes through it. This slower rate of water passing though the bed can create a perched water table that can, in turn, saturate the top layer of the gouge. This saturated zone becomes slick and acts like a lubricant, thus making movement even easier.

A variety of slope failure mechanisms have been observed at Bingham Canyon Mine, such as wedge failures, circular failures, and even debris flow from rain events. But of all these failure mechanisms, the planar sliding failures along bedding planes have been some of the largest and most problematic, even without considering the Manefay. According to Bingham Canyon Mine records of highwall failures, the mine had experienced 10 bedding plane failures from 2007 to 2013 (not including the Manefay), and at least four of them were more than a million tons. The failures were well known and understood, and the geotechnical analysis and planning took into account the weakness of these beds.

However, all of those failures were almost two orders of magnitude smaller in size than the Manefay. Although some of the previous failures developed rapidly, none of them acted in an explosive manner, but instead failed over a longer period of time (hours to days).

Highwall failures or landslides occur when the rock falls into a void. Pit design and construction have a significant effect on the stability of the highwall, especially in relationship to the physical structure of bedding planes and fabric of the rock. If the highwall is oriented in such a way that a bedding plane or other structure dips into the open void of the pit, there is a much greater likelihood that the highwall will fail. However, if the bedding plane or structure is slanted at an angle into the highwall, the potential for a highwall failure is reduced, even if small wedge failures occur along the contact. If the bedding plane or structure is dipping into the pit, gravity is constantly trying to push the rock into the void. Only the strength of the rock keeps it in place, and the cohesion of the surfaces on the bedding plane keep the rock from failing into the void.

The closer the bedding plane or structure is to being perpendicular to the highwall, the less stable the structure and the greater the risk of failure. This makes sense because a bedding plane that is perpendicular to the highwall will have the least distance for falling into the pit. Therefore, it has the least amount of area trying to hold the mass in place, and the high angle means that it has the greatest amount of force from gravity trying to push the mass down into the pit.

The mine design at the Bingham Canyon Mine takes into account the strength, composition, and orientation of bedding planes and structures. Although the pit is basically circular from above, there are a number of notches in the highwall crest that have been designed and excavated to prevent critical bedding planes from failing (such as the Congor, Winnebago, and Fortuna). The purpose of these notches is to change the orientation of the bedding planes and reduce the chance of highwall failure. Figure 6.3 shows the highwall notches in the upper left corner of the photo. The Manefay slide area is shown in the far left-hand side.

Figure 6.3 • Bingham Pit Aerial Photo in 2013, Post-Manefay

Almost all of the geotechnical analyses at the Bingham Canyon Mine had been performed by a geotechnical engineering firm. This consulting firm had extensive experience with the mine because they had been working on the property almost since the consulting company's inception. Over the decades, they had built up their knowledge of the property by providing basic fieldwork, such as mapping of the pit's highwalls, performing lab testing on drill cores, and carrying out geotechnical analyses of the highwalls. Through this work, they developed an in-depth understanding of the mine's geology and structure, its rock types and strengths, as well as the types of failures the mine had experienced.

Through back analysis of those failures, the consulting company also developed an understanding of the root causes of those failures. The failure history database showed that the Bingham Canyon Mine experienced 35 failures between 2007 and 2013, an average of 5 failures per year. Although a majority of these were relatively small in size, the consulting company had an excellent database of information to back-analyze failures. Through this back-analysis work, the consultants were able to determine the typical structures and even the internal fabric of the rock to perform the geotechnical analyses of the mine.

Armed with knowledge of the pit's geology, material properties, structures, and hydrology, the consulting company became adept at modeling and predicting areas that were at high risk of failures. Failures still occurred, but they were not wholly unexpected. The consultants and Bingham Canyon Mine's geotechnical engineers worked together to develop cost-benefit models to determine how far the highwall should be laid back to reduce the number of failures versus the cost of remediating them. Laying back the highwall reduces its overall slope, which makes the chances of failure less likely. However, it is expensive because more overburden has to be moved or less ore mined to achieve the lower-angle highwall. On the flip side, highwall failures can be expensive to remediate in addition to the expense of damaged infrastructure or interrupted production.

The cost-benefit models and analyses by the consulting company did a good job of balancing highwall angles with the amount of potential failures, based on Bingham Canyon Mine's typical failures. An outcome of the cost- benefit models as well as practical experience was that the mine's management expected to deal with failure tonnages from several small failures and not one or two large ones.

Bingham Canyon Slope Failures

Highwall failures are a fact of life in large and geologically complex open pit mines like the Bingham Canyon Mine, which is why the mine experiences multiple highwall failures per year. Most of these failures range from a few thousand to a couple hundred thousand tons, but a few reached a million or more tons in a single event. Considering that the mine had been in existence nearly 110 years before the Manefay failure meant that the mining and geotechnical staff were as experienced and knowledgeable about mining highwall failures as anyone in the industry. Based on that experience and knowledge, the mining engineering and geotechnical staff designed the mine to manage and control the size and number of highwall failures, but eliminating all failures was not possible because of the complexity and variability of the deposits. Eliminating all failures in all circumstances would require designs that reduced the angle of the highwall so greatly that profitable production would not be possible. Instead, the company invested in procedures, processes, and cutting-edge monitoring equipment to detect and predict highwall failures and to protect people and equipment while identifying what was acceptable to the business in terms of failure tonnage and slope cleanup.

The geotechnical analysis methods were developed to predict areas where highwall failures were likely to happen because of geologic conditions. The highwall designs were modified to reduce the likelihood of failure, but there was always some element of risk because of the variability of natural systems like geology. Unlike manufactured materials that can be made within engineered standards, factors such as rock strength have greater statistical variation and therefore are more uncertain.

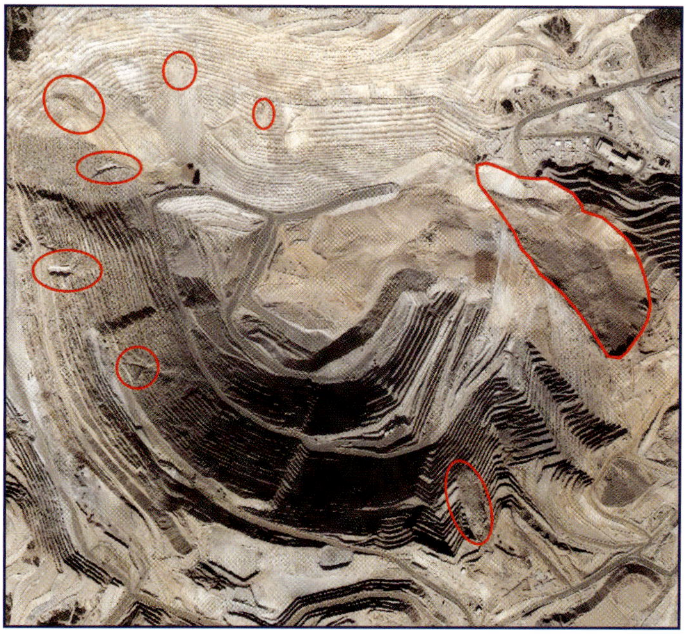

Figure 6.4 • Highwall Failures at Bingham Canyon Mine

With a pit the size of Bingham Canyon, the odds were there would be a few failures that had to be remediated each year. Figure 6.4 shows an overhead plan view of the mine with several highwall failures circled in red. The Manefay failure on the right side of the photo was more than 100 times the size (in tons) than any of the other slides in the photo.

With the introduction of radar monitoring systems, the Geotechnical Team became very proficient at detecting and predicting highwall failures before they happened. Before the Manefay failure, all of the highwall failures fell into the expected range as far as mechanisms and size, which increased the Geotechnical Team's confidence in the analytical tools that were being used to evaluate the stability of the highwall. Most of the analytical work was performed using two-dimensional (2D) models of the highwall and geology that made up the wall of the mine.

Before the Manefay, most of the failures were relatively simple wedge-type failures where the failure mass would slide along a plane of weakness, such as a bedding plane or fault, into the open pit. The size of the failures varied, depending on the angles of the highwall, failure plane, and material type, as well as the amount of water that was present, but all of them failed directly into the pit.

Occasionally, the mine experienced a circular-type failure where there was a zone of weak material. If the material was weak and the highwall too steep, and possibly water present, the highwall could fail by breaking through the rock mass in a circular geometry. Once again, the mechanism was fairly simple and the mass would fail directly into the open pit. Figure 6.5 is a typical cross section of a highwall and shows a circular failure breaking through weak rock shown in red and the wedge failure that slides along a plane of weakness in blue.

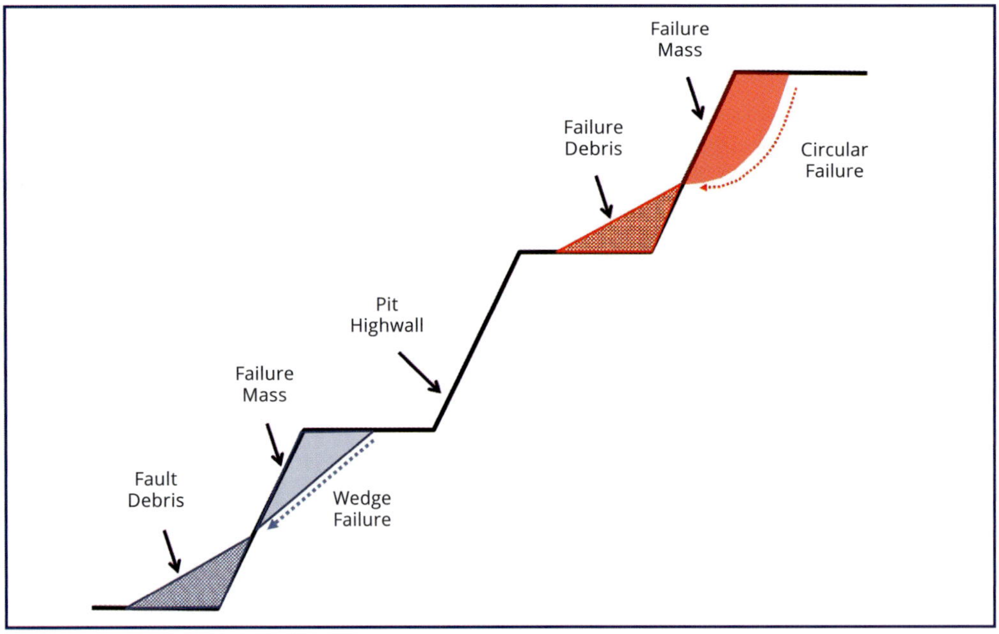

Figure 6.5 • Two-Dimensional Failure Modes

Most of the wedge failures at the Bingham Canyon Mine are controlled by geologic structures, and this was particularly true in the east wall of the mine where the beds, such as the Manefay, Congor, Winnebago, and the Fortuna sill, created the weak gouge surface for the material above it to slide upon. The result of these gouge surfaces is that a majority of the failures on the east highwall were a planar wedge failure, as shown in Figure 6.6. In these types of cases, the wedge would fail along the plane of the weak geologic bed. A second geologic planar feature, such as a fault or joint set, would also need to be present so there would be a side release in the highwall side of the failure.

One of the measurements that the geotechnical engineers measured and designed for was the Bc angle, shown in Figure 6.6. This angle is a measure of how much of the force on the failure mass is going into the highwall versus how much is directed into the open pit. The more that goes into the highwall, the less likely that the mass will fail. Therefore, the smaller the Bc angle, the more stable the mass.

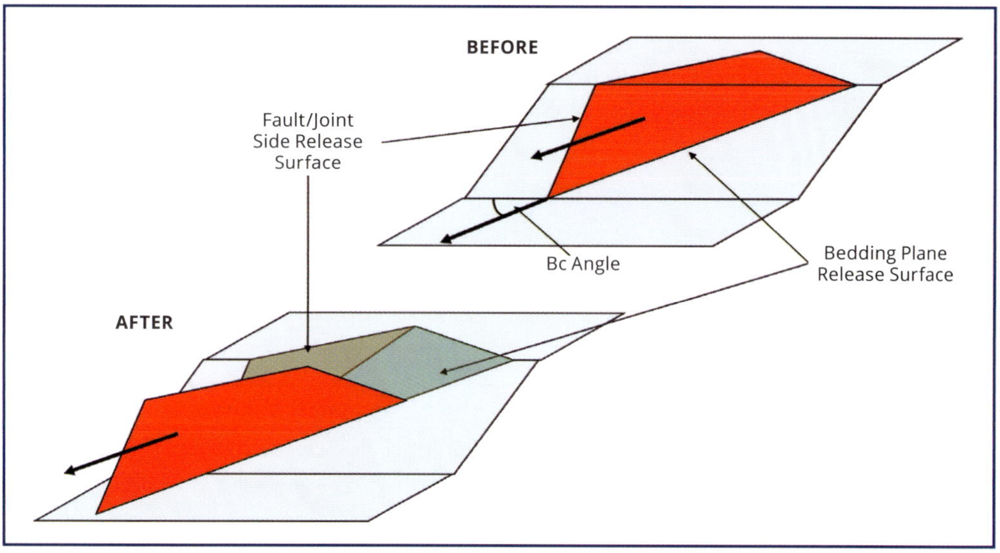

Figure 6.6 • Planar Wedge Failure

The 2D analysis did a very good job of modeling and analyzing both the wedge and circular failures that the mine experienced before the Manefay. The geotechnical engineers and consultants were able to analyze the stability and estimate the odds of failure for the entire pit. The failures did not always give a long period of warning, but they followed a fairly standard pattern of accelerating movement that resulted in progressive failure. When the mass went into progressive failure, it would slide down to a point that it was at the angle of repose (the angle at which granular material will stop slumping because friction is greater than gravity). This process may take hours or even days to complete, and the resulting runout or maximum extent of material movement is based on the frictional properties of the rock mass and amount of water.

The Manefay

Unfortunately, the Manefay was anything but simple. The Bc angle for the Manefay was quite small. Consequently, many of the forces were not moving directly into the pit and therefore should not have been at high risk for a large-scale failure based on historic data. However, the Manefay was not a typical planar wedge failure mechanism, but a much more complex active/passive one. Because of this complexity, which was different than any of the previous highwall failures, the 2D models and analyses were not able to identify and predict the failure before it started accelerating. Fortunately, the mine had state-of-the-art highwall monitoring systems and procedures that detected the impending massive failure well before the event so that people could be protected. Without that monitoring, the outcome could have been very different.

In addition to the failure mechanism being different than previous failures, the fact that the mass was so large, the speed of the failure so great, and drop of the mass so far, the failure did not follow the normal frictional properties for the runout distance. Instead, the mass followed Voellmy properties and acted more like a viscous liquid or an avalanche than a landslide. Consequently, it covered much more of the pit bottom than expected, based on previous failures. This resulted in significant equipment, pit, and infrastructure damage.

As mentioned, the Manefay was a complex failure that could not be identified with standard 2D analyses. Even with 3D analysis, the failure can be difficult to identify without a good understanding of the structural geology. The following sequence of figures and commentary describes the Manefay's failure mechanism, which acted as an active/passive failure. I have estimated the approximate location of the active/passive boundary as well the line between the first and second failure. In addition, the failure mechanism of the second failure is projected by me based on eyewitness accounts and photographs of the failure area that shows the directions of debris movements.

Figure 6.7 shows the general layout and many of the key features of the Manefay on a topographical map of the Bingham Canyon Mine after the failure. The lines of the mine are lines of equal elevation. The closer the lines are to each other, the steeper the topography. The reason the post-Manefay topography is used in this sequence is that it is easier to see the surface that the Manefay mass slid on and therefore better describe the failure mechanism.

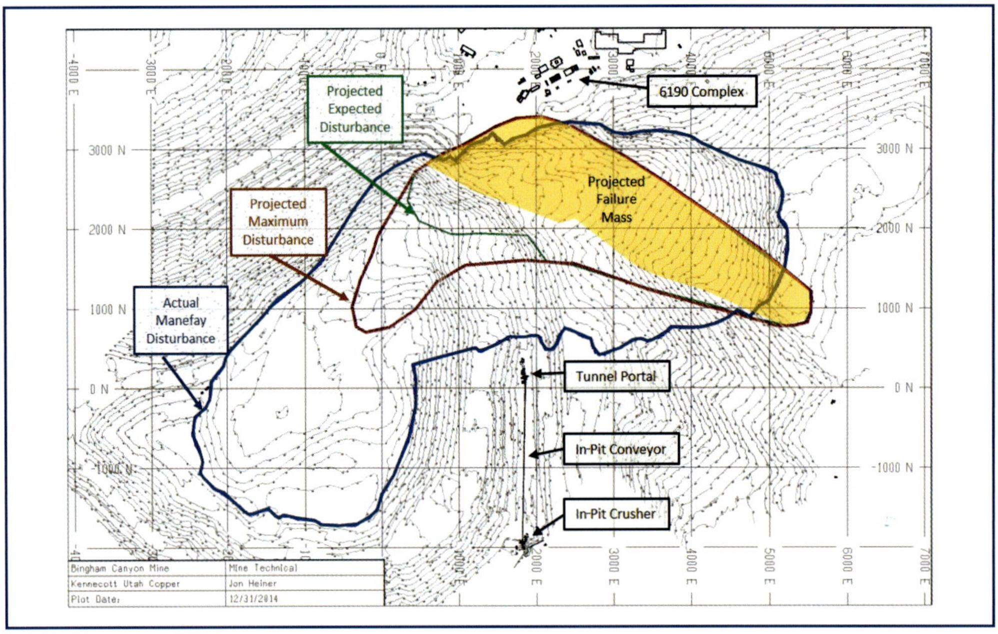

Figure 6.7 • Manefay General Layout

The yellow-shaded area on the map is the projected failure mass that was identified before the failure. The size and shape of this mass was determined with the radar systems that detected movement of the Manefay face as well as inspections by the Geotechnical Department where the engineers had mapped visual cracks on the surface.

The green line—much of which is covered by the brown line—is the expected disturbance area of the Manefay as determined by the geotechnical engineers based on past failures. The brown line is what the geotechnical engineers estimated to be the maximum extent of the failure based on typical frictional properties of the Manefay material. The blue line is the actual extent of the Manefay, which is more than twice the area of the projected maximum. Most of this difference was a consequence of the mass movement having Voellmy (avalanche) properties instead of the typical frictional properties experienced in previous failures.

Figure 6.7 includes the location of the in-pit crusher, conveyor, and tunnel portal as well as the location of the 6190 Complex. The identification labels in Figure 6.7 are not shown in Figures 6.8–6.10 so that the failure mechanism remains clear.

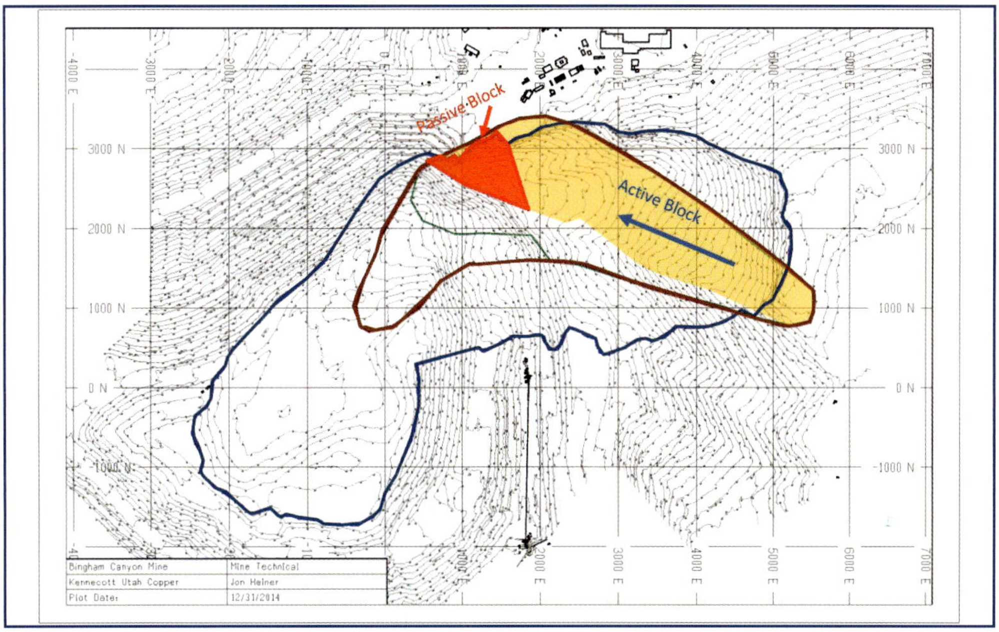

Figure 6.8 • Active and Passive Blocks, Pre-Failure

Before the initial Manefay failure, the entire mass was moving 0.1 inch per day in early February 2013 and was starting to accelerate. The active block was moving down the dip of the Manefay bed, as shown by the blue arrow in Figure 6.8. This movement was being resisted by the massive wall west of the active block, which was solid rock and had no place to move. To a lesser extent, it was being resisted by the passive block that is shaded in red. Unlike the active block, the passive block was dipping slightly into a final wall that was not moveable. The geotechnical engineers could see the movement data and the acceleration from their monitoring, but they could hardly believe that the Manefay would fail. The overall movement was toward the west wall of the mine, which should have prevented the acceleration and failure of the Manefay—but it did not. More than once the geotechnical engineers said the Manefay cannot fail because of the small Bc angle—but the complexity of the failure mechanism dictated a different result.

Between February and April 10, the active block continued to accelerate, which exerted tremendous stress on the passive block. The cracks that had been observed on the surface continued to expand. The geotechnical engineers were closely monitoring the movement of the highwall. The movement was much less significant than the previous smaller failures, but it was the rate of acceleration that alerted the engineers of an impending failure and not the amount of movement per day. On the morning of April 10, the mine was evacuated because the geotechnical engineers believed the failure could happen within a day.

At 9:30 p.m. on April 10, the strength of the passive block was overwhelmed by the weight and stress of the much larger active block, and the passive block failed in a catastrophic manner. Much like squeezing a small seed between your thumb and forefinger, the passive block (shown in red in Figure 6.9) was ejected from between the immovable wall to the west and the overwhelming force of the active block.

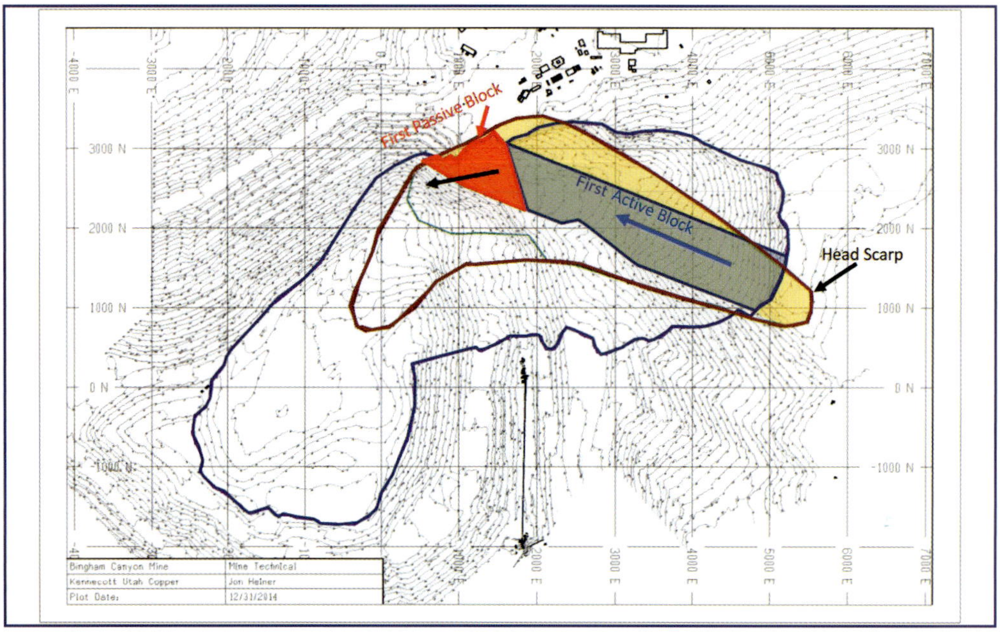

Figure 6.9 • First Manefay Failure

Once the passive block was ejected, creating a sudden change in stress, the large mass of pent-up energy in the active block was activated, and the large mass followed the passive block into the pit at a high rate of speed. Although the actual tonnage of the first failure is not known, it is estimated to be more than 70 million tons. This mass even changed direction as it traveled into the pit, as shown by the black arrow in Figure 6.9. According to seismic data from the College of Mines and Earth Sciences at the University of Utah (Pankow et al. 2014), the entire event took approximately 90 seconds from start to finish. In that time, the mass moved nearly 1½ miles and dropped approximately 2,000 feet. The mass was moving so fast that it flowed more like an avalanche, which was uncharacteristic of the previous landslides at the Bingham Canyon Mine.

The first failure was massive, as shown in the blue outlined area of Figure 6.9, but not all of the active block failed. Some of the block was still being prevented from moving by the immovable wall to the west. The failure left a gigantic scarp that was extremely steep along the entire length of the active block. At the very top of the active block was a piece of the mass that did not fail, which became known as the head scarp. This mass would later become a major concern for the future remediation effort because if it failed, it could put people at risk who might be working in the area below it.

The direction of stress for the active block that was left behind from the first failure was the same as the initial block, as shown on Figure 6.10. The first failure would have left a very tall and almost vertical scarp, which increased the rate that stress built in the remaining block. Because the first block was gone, it changed the area that wanted to fail—creating a new active block that included the gray area shown in Figure 6.10. Most of the first failure was intact rock, but much of the second active block included old mining waste from earlier decades. This material had little strength or cohesion to stabilize the active mass. This new block started to put an incredible amount of stress on the passive block holding it in place.

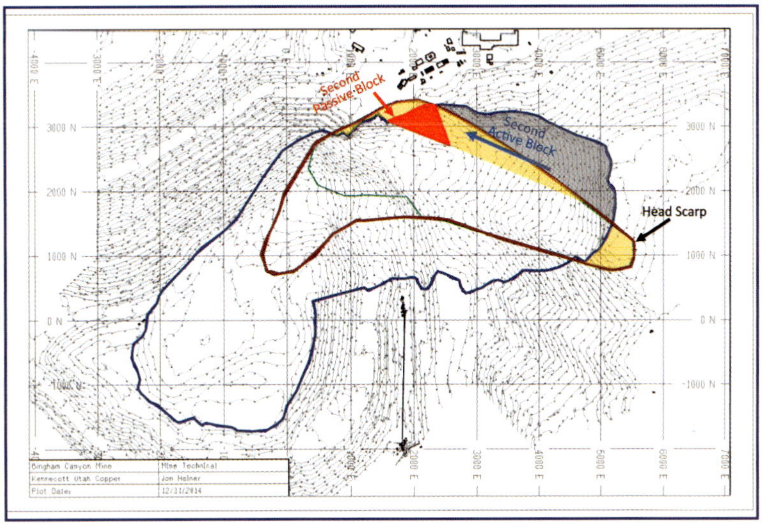

Figure 6.10 • Second Active Block

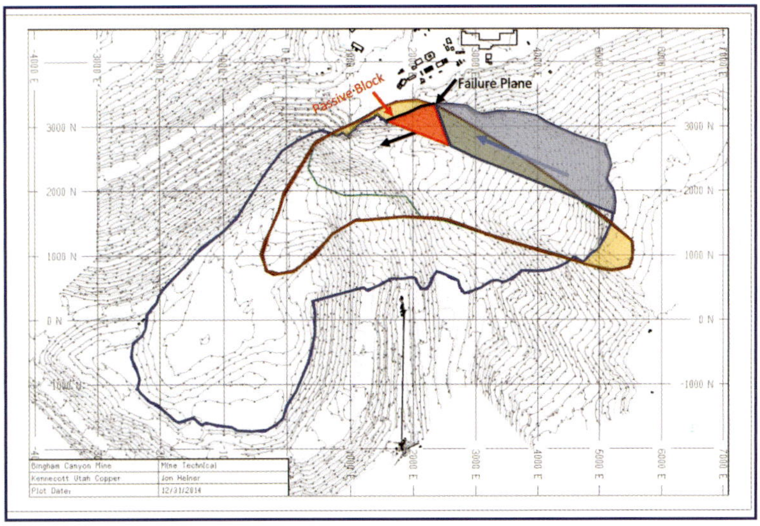

Figure 6.11 • Second Manefay Failure

The stress after the first failure must have been building at an incredible rate because at 11:10 p.m., just an hour and 40 minutes after the first failure, the stress on the second passive block became critical. The stress actually broke the intact rock along the estimated failure plane, approximated in Figure 6.11. The failure generated the sound of a C-R-A-C-K! so loud that eyewitnesses not only heard it but also *felt* it reverberating in their chests.

With the creation of the second failure plane, the second passive block failed en masse and traveled to the open void left by the first failure and toward the open pit. The eye-witnesses stated that the entire skyline disappeared where the second mass once stood. Once again, the new active block, including the activated waste material, followed the smaller mass into the pit. The University of Utah also detected this failure with seismographs, and the second event also took only 90 seconds from start to finish.

In approximately an hour and a half, nearly 144 million tons of material failed in two separate but related episodes. Each event covered approximately the same distance and took about the same amount of time from beginning to end. Both episodes acted according to Voellmy principles, that is, like an avalanche instead of the typical frictional properties experienced in previous highwall failures at the mine. One difference was that the first episode was primarily intact rock, which is typically gray, whereas the second episode had much more previously mined waste, which was brown. Figure 6.12 shows the resulting Manefay debris in the bottom of the pit and along the west highwall where the debris hit and bounced off the wall as it traveled into the pit bottom. It can be seen that the gray material from the first episode creates a ring around the brown material from the second failure.

Even with their extensive experience and background with literally hundreds of highwall failures at the mine, the Bingham Canyon Mine staff and consultants had not considered the possibility of such a large-scale, complex, 3D highwall failure when designing the pit or the Voellmy type of properties for how the debris would flow. However, their world-class monitoring and expertise at predicting highwall failures did prevent what could have been multiple fatalities and infrastructure damage.

Figure 6.12 • Manefay Debris

Lessons Learned Regarding Why the Manefay Failed

Know your greatest risk. Because Bingham Canyon Mine personnel understood that highwall failures were one of the greatest risks to the mine, they placed significant resources and energy into monitoring the highwall. They also used multiple monitoring methods, all of which successfully identified the Manefay risk before it failed in order to protect people from injury or death.

Try to determine the unknowns. In large and complex processes that deal with natural systems, such as in mining, there will always be some level of risk. Having as much data as reasonably possible reduces the risk of unknowns. In the case of the Manefay, there were limited data regarding water or rock strength on the bed of the Manefay. The back analysis that was done after the failure indicated that the strength of the bed was lower than expected. Additional drilling and data might have improved the geotechnical analysis of the area.

Beware of overconfidence and complacency. The Bingham Canyon Mine team and its consultants had significant experience and success with understanding historic failures and predicting and managing failures at the mine. This led to a high level of confidence in their capabilities as well as some hesitation to utilize outside resources that may have had less knowledge yet a different perspective. This was exacerbated by the critical nature of monitoring the Manefay movement and not wanting to take time away from that work to bring outside people up to speed on the situation.

Question changing situations. The Geotechnical Team did a tremendous job of identifying the Manefay. However, the Manefay was nearly two orders of magnitude larger than any failure at the mine. That great of a difference should trigger questions about what is known and believed. All data and assumptions should be reviewed and reconsidered when situations change.

Encourage independent/external review. Independent and/or external review of critical risks and situations should be standard procedure to challenge current beliefs and prevent isolation of perspectives and ensure that alternatives have been considered. It is possible that a non-mining geotechnical expert would have considered Voellmy-type properties to be a greater possibility based on other large geologic failures instead of just considering mining highwall failures that are normally frictional type failures. Inviting someone with a different perspective to review work does not guarantee that all possibilities will be considered, but at least there is some possibility that more information will be revealed.

Innovation, Technology and Culture Change

Chapter 7

The Manefay recovery work was the result of innovations and technologies that were implemented by the employees working in a culture that empowered them to meet demanding challenges. It was a real testament to the Kennecott employees, vendors, and consultants who applied technologies in novel methods to meet what were thought to be impossible goals.

People went above and beyond traditional thinking to use existing technology in innovative ways to keep people safe in the recovery effort, maximize copper production in the short term, prevent a future gap in ore production, and maintain downstream operations. So much was happening during the recovery that this chapter could be a book in itself, but this discussion provides a sample of the great work that was done after the slide.

Nontraditional mining equipment was brought in and remediation efforts focused not only on the pit, but on downstream operations as well. Remote-control capabilities were installed on equipment, haul truck operators were trained to use dozers, monitoring problems were solved with sporting goods, a team was built to ensure independent review of mining activities, and the company supported a positive cultural change. Lessons gleaned from this chapter focus on creating conditions for innovation, performing independent reviews, creating positive cultures, and understanding the needs of the entire organization.

DownStream Operations

While the mine was struggling to recover from the Manefay failure, the downstream plants had a host of their own problems to solve. Large industrial processing plants like the concentrator and smelter benefit from a constant flow rate of consistently good-quality material to run efficiently and cost-effectively. After the Manefay, the mine was no longer able to supply that constant supply of material until it could return to normal operations. One of the immediate effects of the Manefay was that for the first two and a half weeks after the event, there was no ore available to be delivered from the mine to the downstream operations. Even after the pit ore mining operations resumed, ore production rates would be reduced and would not meet the needs of the concentrator and smelter.

Low-Grade Stockpiles

The mine had built stockpiles of low-grade ore on the outside portion of the pit to supplement production during times of reduced ore production from the pit. Under normal operations when the stockpile was used, the low-grade material would be loaded into haul trucks and transported back into the pit so it could be crushed and sent on the conveyor to the A-frame at the Copperton Concentrator. The A-frame is a large covered stockpile of crushed ore that is used to feed the grinding mills in the concentrator. Before the Manefay, there was no established method to crush or transport material from the low-grade stockpile directly to the A-frame. Figure 7.1 shows the general arrangement of the low-grade stockpile in relation to the Bingham Canyon Mine and Copperton Concentrator.

Not only would the team need to transport the material to the A-frame, they also had to find a way to prevent large rocks from plugging the chutes under it. Normally, the low-grade ore would be sent through the in-pit crusher, where it would have been crushed to 10 inches or less. Installing a large crusher like the one used in the pit to reduce the low-grade stockpile material would be costly and time-consuming to acquire and build. The team instead opted to rent portable screens and a small, moveable crusher used in sand and gravel pits to remove oversize material. Even though this plan required handling the low-grade stockpile material twice—first, loading it from the stockpile to the screen, and then a second time loading it into the 40-ton trucks to be transported to the A-frame—this solution would prevent boulders from damaging equipment or stopping operations.

Figure 7.1 • Low-Grade Stockpile Location

Before the Manefay failure, a team of engineers under the direction of the mine's previous general manager had studied ways to move the low-grade stockpile to the A-frame. Meeting that goal required the construction of nearly six miles of road from the low-grade stockpile to the A-frame at the Copperton Concentrator. This road would be expensive and time-consuming. The idea was not immediately implemented because the mine planned to return to full operation quickly after the failure. It was believed that the only real risk of delayed ore production was if the slide took days or weeks to settle. Ultimately, the real impacts of the Manefay were much greater than anticipated, and the day after the slide it was realized that having the ability to haul the low-grade stockpile material to the A-frame could be beneficial to providing feed to the concentrator and ultimately producing additional copper.

The low-grade stockpile production could only meet 10% of the downstream plant's throughput requirements, so the concentrator would not be able to operate at full capacity. In addition, the low-grade stockpile material contained only one-third of the copper per ton as normal run-of-mine ore, so the smelter would be starved for concentrate. But at the time, any amount of ore would improve the downstream operations.

In the immediate days after the failure, the team quickly finalized plans to deliver the ore from the low-grade stockpile to the A-frame using two roads, the first being a temporary road for smaller trucks that would be quickly dozed and used until a second, more permanent road was built for the large haul trucks. The first temporary road (perhaps the word *trail* is more accurate) would be relatively narrow, curvy, and steep in sections because of a drop of more than 800 feet in elevation. Since this temporary road would not accommodate the mine's 240- or 320-ton haul trucks, the team would need a fleet of 40-ton articulated trucks designed to maneuver around tight corners. Because the equipment and operator skills needed were unlike typical mining operations at the mine, the team contracted with Granite Construction, Ames Construction, and W.W. Clyde for both the Caterpillar 740B and Komatsu HM 400 trucks and skilled operators.

Figure 7.2 • 40-Ton Articulated Trucks Dumping at the A-Frame

Figure 7.3 • Low-Grade Haul Road

The planning for the temporary low-grade stockpile road was completed on April 14, just four days after the Manefay slide. The work started immediately, and limited haulage of the stockpile material to the A-frame commenced on April 16. This was a terrific accomplishment for such a short period of time, which can be largely attributed to the tremendous cooperation and support from the engineering team, contractors, Rio Tinto Procurement, Rio Tinto Technical Services, and the mine's Operations people. Figure 7.2 shows 40-ton haul trucks dumping at the A-frame. After the low-grade ore was dumped, dozers would push the material to the drawpoints so it could be loaded onto conveyors and transported to the concentrator.

The volumes were not large—starting out at less than 7,000 tons per day and building up to 25,000 tons per day—but it was the inaugural production of copper after the Manefay.

Using the 40-ton articulated trucks was only a temporary solution for low-grade haulage. Design work commenced immediately on the permanent haul road. By May 13 construction began, and by July 17 the 6-mile road was complete. The new road was straighter, wider, and had lower grades to accommodate the 240-ton trucks. Dust was easier to control with water trucks and better road material. The daily production increase was marginal because the screens were a limiting factor, but the road was now safer to drive on. Figure 7.3 shows the location of the newly constructed low-grade haul road.

The haulage of the low-grade stockpile contributed approximately 3.8 million tons of ore to production, representing about 15% of the total material sent to the concentrator during the recovery of the Manefay. The innovative team found a way to use nontraditional equipment and outside operators to quickly move the low-grade material into production until a longer-term solution could be implemented. The team demonstrated vast ingenuity and a high degree of teamwork among employees and vendors to achieve what they did in the short time frame.

High-Silica Rates

The normal feed rates from the mine are limited by the amount of ore the concentrator can process, which in turn is dependent on the hardness of the ore. Typically, the crushing and grinding systems that the concentrator can process range from 140,000 to 170,000 tons per day. During late April and May of 2013, the mill received approximately 93,000 tons per day of ore from the mine and another 20,000 tons per day of the low- grade stockpile. This was well below the crusher and grinding capacity, especially considering that the low-grade stockpile was much softer than the pit ore, so the system could have been able to exceed its normal rates.

The 68 shovel and one loader were doing an amazing job of providing more than half the normal mill requirements, but the production rates were not constant. Anytime the 68 shovel did not operate, pit production almost came to a halt. In addition to the variable rates, the Manefay debris covered up more than half of the ore production faces that would normally be mined at any one time. Being able to mine in multiple ore sectors is beneficial to the downstream plants for blending purposes. The smelter is sensitive to the blending because significant variances in elements such as silica, or a ratio of elements such as copper to iron, can influence the throughput and effectiveness of the operation.

The smelter was also affected because it is designed to operate within specific production rates. If the rates go too low, the process breaks down and the smelter has to be shut down. The smelter operations try to avoid shutdowns if at all possible because the cooling and subsequent reheating can damage or reduce the life of the equipment. In the time after the Manefay, the smelter was affected by problems with both quality and feed rates. All of the ore production in the pit was coming from the E4 sector, given that the other areas were covered by Manefay debris. The E4 sector had silica content that at times was 20% greater than normal. A silica percentage above 18% could not be smelted because it would permanently damage the linings within the furnace and potentially lead to a catastrophic failure.

To solve the high-silica issue, the smelter increased the stockpile levels of concentrate (the feed material to be smelted) so that lower silica concentrate could be used to blend out the high-silica material. The stockpiling of concentrate compounded the problem of low production quantities, and the smelter was in danger of going below the minimum throughput levels. A decision was made in June to actually shut down the smelter for two weeks to increase the inventory levels. By temporarily shutting down, it could work at a more constant throughput level when the smelter started back up and would have enough different grades of concentrate so the high-silica material could be blended out without destroying the furnaces.

The Rio Tinto Marketing group searched the copper industry to purchase concentrate to supplement the ore from the mine. In particular, they were looking for low-silica concentrate, which would help with the blending of the silica as well as increase the throughput rates and keep the smelter operational. The problem was that Kennecott's Garfield smelter had always been dedicated to the Bingham Canyon Mine and did not have an established method to bring in large quantities of copper concentrate. Fortunately, there was a rail line into the smelter complex, but there were no unloading facilities.

By April 24, the Marketing Team located a source of concentrate from a copper mine in Nevada. The concentrate would not be delivered until early May, so the Smelter Team made plans to unload the train cars that would transport the concentrate. A simple method of using a small backhoe to unload the cars was deployed, and soon the delivery of purchased concentrate had begun. There were limitations to the amount of purchased concentrate available that met the blending requirement as well as limitations in the unloading capabilities; consequently, purchased concentrates made up less than 10% of the overall smelter requirements. But the purchase of concentrate made a large difference in the smelter being able to blend the highly variable ore from the mine as well as maintain a more constant feed rate through the smelter. Figure 7.4 is a photo from the top of the smelter stack and shows one of the concentrate trains coming into the smelter yard (top right-hand side) with the backhoe waiting to unload the railcars (bottom right arrow).

Figure 7.4 • Concentrate Train at Smelter

Cost Reduction

The next problem for the smelter was to reduce its costs. Similar to most smelters, a significant portion of its costs were considered fixed, in that they did not increase or decrease with production. The largest of these fixed costs before the Manefay was labor costs. The Manefay had a large effect on the throughput of the smelter and, in the early months of production, was nearly cut in half, which meant that the fixed-cost portion of the smelter doubled on the basis of cost per pound of copper produced. The company had a tremendous financial problem: costs at the mine increased so it could recover to full production, but in the meantime, the copper production had significantly diminished. Several methods were employed to reduce costs, such as the creation of Mandate Teams to optimize the operation. However, even for these efforts to be successful, the fixed cost of the downstream plants and overhead had to be reduced somehow, which meant that the number of people would also have to be decreased.

Reducing the number of employees is always a difficult task, but the Senior Leadership Team set about doing so by following their core principles as closely as they could. The desire of the leadership team was to reduce the workforce as much as possible through voluntary means instead of a layoff, yet also retain employees who had the right skills and the greatest motivation. In addition, the company also had to stay within the confines of the Kennecott Collective Bargaining Agreement.

By mid-May, the company knew that approximately 30% of the positions downstream from the mine had to be eliminated, mostly in the smelter and concentrator area. The Kennecott Collective Bargaining Agreement did not allow for the company to offer incentives for employees to retire. The Senior Leadership Team developed a plan to provide incentives for employees to resign if they met specific requirements. This plan was shared with the union leadership before being presented to the workforce, and there was agreement that it was a fair and equitable plan.

The request for volunteers to resign went out to the workforce in hopes that a layoff could be averted. The response was at least as good as expected, and the layoff of employees could indeed be avoided. Unfortunately, several of the people who were leaving Kennecott worked in areas of the company that did not require a reduction in personnel, such as mining operations. The managers and superintendents of each of the departments worked together to identify people who would have to be transferred to other areas of the company based on the contractual layoff criteria. A plan was put in place and on June 3, a supervisor met with each of the employees who would have to be transferred. Each was given an option to take a layoff or transfer to another part of the company that needed additional people. When the day was over, almost every employee on the transfer list accepted his or her new position. The company was able to achieve a major reduction in force and did not have to lay off anyone.

This reduction demonstrated the company's commitment to treating employees as fairly as possible. Everyone understood that changes would have to happen with the reduced copper production, but the way it was handled actually built on the trust that the company had fostered from the communications before and after the Manefay.

After the reduction in force, the smelter continued to find ways to reduce costs. The team established a Cost War Room where team members tracked both the costs and the projects to reduce the costs. Cost cutting became a way of life at the smelter, and employees at all levels contributed to finding ways to save costs. By the November time frame when the mine returned to full production, the Smelter Team was highly successful in cutting costs—to the point that the cost per ton did not increase at the smelter but actually decreased—even though there was a significant reduction in copper throughput. This was a remarkable achievement for the men and women working at the smelter.

Copperton Concentrator

The Copperton Concentrator plant also realized a significant reduction in tons of material being delivered to the plant at the same order of magnitude as the smelter. Unlike the smelter, the Recovery Team was focused on getting as much copper from the ore, as much or more than cutting costs. The production of the low-grade stockpile mitigated to some degree the loss of ore tons the concentrator had to process in a day. However, the copper grade of the low-grade stockpile contained one-third of the copper per ton as compared to the ore that was being mined at the Bingham Canyon Mine. As a result, the concentrate production from the plant was down significantly from the original plan. Therefore, it was critical that the Recovery Team recover as much copper as possible from the ore.

The Recovery Team was given a goal to increase recovery by 5% above the percentage designated in the original mine plan. This was a difficult goal given that the low-grade stockpile material generally achieved a lower recovery than expected because the material had been oxidized over time, which had a negative effect on the overall process.

The team focused on improving their grinding and flotation performance. In particular, they took time to improve the process conditions on the rougher-scavenger to help improve recovery. The improvements were dramatic. The Recovery Team exceeded its goal and actually achieved a 6% recovery increase.

Lean Methods

All of the sites (mine, concentrator, smelter, and refinery) leveraged off the entire workforce to find ways to improve and meet their goals. They used Lean methods and technology as the foundation of the business improvement process. Anna Wiley, the general manager of business improvement and reliability, was the greatest advocate for this program and the driving force for making it work. Although a variety of tools were employed during this time, three in particular were used across all of the sites: information center boards, mandates, and kaizens.

The *information center boards* are bulletin boards set up in strategic locations around the mine site. Although the boards were meant to inform employees of what was going on, their real purpose was to be a place where teams could meet each morning to briefly discuss how the progress of their projects was going, what needed to be done next, and who would do it. The idea was to have meetings that lasted 15 minutes or less and were held in locations where all members had to stand (no chairs allowed) to help ensure that the meetings did not take too long. The information center boards were used by a wide variety of groups, such as the Mine Management Team, Technical Services Team, and the Next Ore Team. Each group that had an information center board generally had a Lean advisor helping to set up the board and assisting the leader of the group to effectively use the tool.

Mandate Teams are large project teams who gather to solve critical business problems that are most important to the success of the company, as mandated by management. The mine had a series of mandates that resulted in the formation of the First Ore Team, Next Ore Team, and the E5 Team. The smelter had the Cost Team and the concentrator had the Recovery Team. These teams had full-time multidisciplinary members that included a manager, superintendents, supervisors, operators, and technical staff. As a general rule, no one could say "no" to a Mandate Team's request since that was the most important work going on at the time.

Kaizen is the Japanese word for "improvement," and kaizen events are smaller groups that work on a smaller and more focused event, such as reducing the number of minutes to perform a task within some part of the operation. These events may be part of a mandate, but could also focus on any part of the business that needs to be improved. For instance, there was a kaizen for reducing the haul-truck cycle time on the 10% haul road because of the congestion for the E5 mandate. In another case, there was a kaizen to reduce the amount of time that haul trucks spent on the dumps in the Cornerstone Operations at the mine. Although not a part of a mandate, it was an important time-saving measure to make the haul trucks more productive.

What was interesting after the Manefay was the acceptance—if not the excitement—of individuals in using the Lean tools. People almost religiously showed up at the information center boards, because it was the place they could quickly find out what was happening. They also did not want their name or group coming up as a reason that the company was not meeting its goals, so they made sure their part of the project was complete.

Employees at all levels were also proud to be on a Mandate Team, especially right after the Manefay. The mandates were there to actually save the company, in their view, and only took the best. So many people believed that if they were on a Mandate Team, they were part of a select group working on the most important work. Over time, some of the glamour of the mandates wore off. Being on a Mandate Team meant a fast pace, long hours, and high stress, which most people handled well, unless they were on multiple mandates in a row.

The kaizen events involved employees who were primarily from the hourly workforce. These groups also benefited from very high participation after the Manefay. There was a general feeling that the group members could get their good ideas and solutions implemented—not always an easy task in a mine. The kaizen groups also received a lot of recommendations. The operations were rightfully proud of the changes that the kaizen members were able to make, so when almost any group from outside the company would visit, there would be a presentation by one or more kaizen-event groups. Overall, the kaizen members were successful in making changes and were also recognized for making those improvements.

Remote-Controlled and Autonomous Equipment

Some of the most innovative work after the Manefay was the acquisition, implementation, and use of remote-controlled and autonomous equipment. Before the Manefay failure, a remote-controlled dozer and front-end loader were used at the Bingham Canyon Mine to work in hazardous areas in the rare occasion when manned equipment could not be used safely. The recovery from the Manefay would require the use of a large number of both remote-controlled and autonomous equipment to work on and below the dangerous and unstable 600-foot-tall scarps. The work would require several dozers, excavators, loaders, and drills working in very challenging environments. There was also the potential need for a shovel and haul truck. It was obvious that the Bingham Canyon Mine would have to go from an occasional use of remote-controlled pieces of equipment to one of the largest users of innovative and technical equipment in the world, and it would have to happen quickly.

Figure 7.5 • Atlas Copco Pit Viper 275 Drill

In all, remote- and tele-remote-control capabilities were installed in 32 pieces of equipment. Two additional drills were upgraded to full automation. Just like a remote-controlled toy, the remote-controlled equipment was operated by a person with a radio controller. The autonomous equipment was programmed and required no human intervention to perform once operation was initiated. The remote-controlled and autonomous systems were supplied by seven manufacturers. Table 7.1 lists the equipment that had remote-control and autonomous technology installed to help remediate the Manefay landslide.

Figure 7.6 • Tele-Remote System for Atlas Copco Drill

Because of the large number of systems that were required, multiple companies were used. Each company had different products, areas of expertise, or capabilities and all of these would be required to meet the needs of the remediation work in the time frame required. Additionally, the timing requirements of the remediation work meant that a large number of resources would be needed to get the work done quickly; the only way to obtain all of these resources was to use all of the available companies. Figure 7.5 shows the drill just after commissioning on August 19, 2013. This drill included tele-remote capabilities so it could be operated remotely from the cab of a pickup, as shown in Figure 7.6. In a tele-remote system, the operator panel has the same look, feel, and buttons as the actual equipment.

Table 7.1 Remote-Controlled Equipment

Units	Equipment	Models	Remote Manufacturer/System
3	Dozer	D8T	Remote Control Technologies
10	Dozer	D10T	Caterpillar/Command
5	Dozer	D11T	Caterpillar/Command
4	Excavator	Cat 374	Autonomous Solutions Inc.
2	Excavator	Cat 390	Autonomous Solutions Inc.
1	Shovel	Hitachi EX 5600	TORC Robotics, LLC
1	Haul truck	Cat 793D	TORC Robotics, LLC
2	Loader	Cat 992G	CAST Resource Equipment/Cattron-Theimeg
1	Drill	Atlas Copco Pit Viper 271	Atlas Copco/Tele-Remote
1	Drill	Atlas Copco Pit Viper 275	Atlas Copco/Tele-Remote
1	Drill	Atlas Copco DML	Flanders Electric/Ardvarc (full autonomous)
1	Drill	Sandvik D90	Flanders Electric/Ardvarc (full autonomous)

Systems by Caterpillar, Atlas Copco, and Remote Control Technologies were off-the-shelf systems that had previously been developed and were available on the open market. These systems were tried and tested, so it was a matter of getting them delivered—or, in some cases, manufactured and then delivered—as quickly as possible. The systems supplied by Autonomous Solutions Inc. (ASI), TORC Robotics, and Flanders Electric were different in that they were not existing systems that could simply be ordered and shipped. They had to be either modified from existing designs or developed specifically for the equipment needed for the remediation work. Figure 7.7 shows one of the new Caterpillar excavators being operated remotely with a system by ASI.

Figure 7.7 • Cat 374D Excavator with ASI Remote Control

The system by CAST Resource Equipment used an off-the-shelf controller by Cattron-Theimeg. This system was already on a Cat 992 loader at the mine and was not new for the mine, but the system was used infrequently. Functional changes were quickly made on the loader's breaking system to improve productivity in the remediation work.

From the start, it was a challenge to implement the large number of remote-control and autonomous systems in the short time period required. The fact that there were so many manufacturers and several of the systems did not actually exist or needed modifications made the task seemingly impossible. Fortunately for the Next Ore Team, the Underground Team at Kennecott had a contractor working for them who was an expert on remote-control and autonomous systems. Mark Sauder was an independent contractor who specialized in unmanned aircraft and ground vehicles. He had been previously engaged by Kennecott, so he brought an in-depth understanding of the technology required for fielding these types of systems.

Mark was integrated into the Production Support Team, reporting to David Olson and Jessica Sutherlin, with the responsibility of getting the remote-control systems operational as quickly as possible. With support from his supervisors and Jessica's team of specialists, including Eric Cannon and Chris Haecker, Mark helped identify suppliers and available technologies. He then reviewed and ensured that all design specifications were met for the systems under development. Equipment, such as the excavators, was needed as quickly as possible, so the team worked closely with each supplier to develop systems. Dozers were ready in the first two weeks after the Manefay, and the first excavator systems were ready the first week of June—which was impressive because these were the first ever builds for that class of equipment. The systems that were needed for cleaning the benches below the head scarp and demolishing the Bingham Shop were online in record time. Every machine built was critical for work to be completed in a safe manner. The team then worked hard to make sure the operators felt comfortable with the remote-control technology.

During the ensuing months, all of the companies working on the remote-control and autonomous systems did amazing work. The results of the large number of systems installed by Caterpillar, the autonomous drilling capabilities designed by Flanders Electric, the responsiveness of Remote Control Technologies and Atlas Copco, as well as the innovations of remote-control systems by TORC Robotics, ASI, and CAST, were more than impressive—they were amazing. The one common denominator in bringing about the operational capability of each of these machines was Mark and his team.

Without the help and support of each of these manufacturers and consultants, Kennecott would not have been able to operate as safely and effectively as they did after the Manefay; these companies and specialists rose to the occasion and made a difference.

Using Remote-Control Systems

After the remote-control systems were in place, the next step was to make that equipment effective. Operators of equipment such as dozers and excavators use their sense of sight, sound, and feel as well as their understanding of the controls to be effective. With the remote-control systems, the operators could operate the machinery, but it was difficult to be productive when physically located away from the equipment and not having the perspective (and comfort) of being in the cab.

The team tried several procedures to improve the effectiveness of the remote-control operators. When possible, the operators would work from the cab of a pickup truck, but when they needed a better view, they would stand outside for hours at a time in the glaring sun and wind. A simple solution was to purchase canvas chairs with umbrellas, as shown in Figure 7.8. The operators were teased that they were out on a picnic, but in reality, it was hot and tiring work because of the concentration required to operate equipment remotely.

As the work progressed, the operators would have to work farther and farther away from the dozers. One of the operators determined that the best vantage point to see the dozers was from above. An innovative solution was to rent bucket trucks, the type used to work on power lines. The operators were now 50 feet in the air and did have a better view of the dozer, but they were also subjected to even more of the wind that rocked the bucket. Umbrellas again were used to provide shade, but the gusts would blow away the umbrellas. There were many trips to the store to replace the umbrellas! Figure 7.9 shows two operators controlling a dozer and an excavator from a bucket truck.

Figure 7.8 • Remote-Controlled Dozer

Figure 7.9 • Remote-Control Operations

As work continued on the bench cleaning, the operators would be nearly a thousand feet away from the equipment and did not have a good line of sight. In those cases, cameras were mounted on the equipment and the video was transmitted to monitors in pickup trucks parked more than 2,600 feet away. These operators worked from the relative comfort of the pickups.

In early 2014, TORC Robotics took this concept even further when they built remote-control systems for the Hitachi shovel and Caterpillar haul truck. Multiple video feeds from the equipment were sent to an equipment simulator. For the operator, it was like training in the simulator, but he was actually operating the equipment remotely.

Training Dozer and Excavator Operators

When the Next Ore Team was formed in early May, the team realized that a large bottleneck to remediating the Manefay was not having an adequate number of qualified operators for the dozers and excavators. The Production Support and Procurement Teams were busy acquiring 35 dozers and 13 excavators, in addition to 32 remote-control systems, all of which would require qualified operators.

The Bingham Canyon Mine is a huge operation, with many excellent equipment operators. Some of these operators were already experts at performing the remediation work from their experience of working on multiple smaller highwall failures. But the Manefay remediation was one to two orders of magnitude larger than any previous failure at the mine, and more than 150 dozer operators would be needed for the short-term day-to-day operations and remediation work.

The decision was made to use experienced dozer operators from the workforce and train additional operators. Most of the people to be trained were haul truck drivers because they constituted the largest number of equipment operators in the company. Karen Bakken, a young mining engineer from the Next Ore Team, was assigned to coordinate and manage the training of these individuals to operate the remediation equipment.

Training 150 people to operate a large piece of equipment like a D11 dozer in a short period of time seemed to be an impossible task. The first thing that Karen and the Training Team did was survey the workforce to find operators who had previous dozer experience. The team identified 60 employees who had this experience and were not already operating dozers on a daily basis. Two expert dozer operators were selected to test these operators to determine at what level they could perform. Kennecott had high standards for equipment operators, and people would be trained until they met those standards.

Of the 60 operators with dozer experience, nearly 40 of them required only minimal additional training. Training started with the most experienced and highly rated operators. Through the process, they developed a training, testing, and sign-off procedure that was specific to the hazards and methods of remediation work. After the process was formalized, they realized they would need more trainers to train the larger number of inexperienced operators. Kiewit, a construction and mining company with a branch office outside of Salt Lake City, was selected to supply additional trainers because of their experience managing large-scale reclamation projects that required many dozer operators.

Nearly 60 of the next set of operators selected had little to no experience with operating a dozer. The training process was intensive, and no shortcuts would be made to just push people through. Consequently, training was scheduled for 24 hours a day, seven days a week. The selection process and logistics of keeping this process going was intense, but Karen kept everything going forward.

Between May 20 and July 7, the Next Ore Team prepared 110 operators to operate dozers for the remediation work. Some of the operators were also trained to use the remote-control systems. It was an amazing site to watch the equipment operating in the training area without an operator in the cab. As soon as the operators were safe and proficient with operating the dozers, they were tested. If they passed the test, they would be signed off by the testers and then sent to the remediation operations. Figure 7.10 shows an operator using a remote-controlled Caterpillar D8T dozer in the remote-control training area.

Figure 7.10 • Training on a Remote-Controlled Dozer

By late June, the training of the dozer operators was beginning to have a major effect on the Cornerstone operations. The priorities of the company were safety and ore production, so operators were assigned to equipment for remediation efforts and mining the bottom of the pit. Because Cornerstone operations were stripping waste in the upper part of the pit, equipment for that project was idled because there were not enough operators. By the time 110 dozers operators were trained for the remediation work, a major part of the haul truck fleet was shut down and the Cornerstone operations were falling behind in waste production, placing the future of the mine production in jeopardy.

Before the Manefay failure, two shovels and a number of trucks were transferred from the bottom of the pit to Cornerstone operations because the Manefay failure would destroy a large portion of the 10% haul road. The production from Cornerstone was relatively high shortly after the Manefay because of the additional equipment, but as each additional truck driver was trained as a dozer operator, a haul truck would be shut down for a corresponding number of hours.

By early July it was obvious that additional operators would be required to keep all of the production and remediation equipment operating. The remediation work would be on a short-term basis, so contract operators were brought in to supplement the Bingham Canyon Mine's workforce. Although the contracted operators supplied by Kiewit were experienced, they still had to complete the same testing, site-specific training (if needed), and sign-off that the Kennecott employees had to go through. Both the Kennecott and Kiewit dozer operators did an excellent job completing remediation work and were instrumental in the speed of the recovery.

Having sufficient qualified operators was just as critical to the success of the remediation work as having enough equipment. The work that the Next Ore Team did to find and train the operators ensured that there were an adequate number of operators to complete the work safely. This is great example of how planning for details made a difference in the successful recovery from the Manefay that could have easily been a major bottleneck.

North Wall Drainage Gallery

The north wall drainage gallery was a set of tunnels being built to dewater the northwest side of the mine. After the Manefay slide, the Underground Team was restricted from entering the tunnels to check the water levels in the underground workings. If the water level was too high, the underground workings would flood, causing significant damage to the workings and equipment. If the water level was too low, the pumps would overheat and burn out—and worse, this would lead to future flooding depending how long it took before the pumps could be replaced.

To measure the water level and determine when the pumps should be turned off, the Underground Team hired a contractor who used a special camera that was dropped down the drill holes to visually inspect water levels. Unfortunately, the contractor had problems with the camera equipment and could not get it down the hole. After several attempts, the contractor decided it was best not to risk the expensive equipment and told the team it could not be done.

The team members were stuck because they did not know whether to turn the pumps off or leave them on. The longer they waited, the greater the chance there would be a problem. At that point, Jeff Dunn, the hydrologist for the Underground Team, and Ian Schofield, with the Environmental Water Quality group, came up with an ingenious idea: they would build their own camera to check the water level. They drove to town and purchased a GoPro camera, green light sticks, and a spool of cable.

When Jeff and Ian returned, they attached the GoPro to the cable. Members of the Underground Team stopped by to see what they were doing and expressed their amusement for trying such a hare-brained idea. When they were ready to lower the camera, they bent the green light sticks to initiate their glow properties and dropped them into the shaft, thinking that the light sticks would float at the top of the water level. More members of the Underground Team came by and laughed outright when they saw them throwing the light sticks down the shaft. Jeff and Ian did not throw just a few light sticks down the shaft—they probably threw down a hundred of them. The two guys were the talk of the town and everyone had a good chuckle.

The next morning, Jeff and Ian shared the GoPro video with the rest of the team. As the camera was lowered down the shaft, the video revealed a strange green glow that got brighter and brighter the further down the camera went. By the time it reached the water level, it was bright enough that the miners could see the walls of the shaft in the video—and they were able to determine the level of the water!

The rest of the Underground Team were no longer chortling at the "nutty" (and innovative) idea of a GoPro and light sticks. In fact, when they saw the water level in the video, they actually stood up and cheered! Miners are not known for showing a lot of emotion, but they also realized that Jeff and Ian had solved a problem that prevented a significant amount of work in the future.

Independent Review

The fact that the Manefay did not act as expected and had destroyed a significant amount of equipment demonstrated the need to look at the geotechnical analysis and planning differently than what was done before the Manefay. In addition to the need for rethinking the previous work, the mine had to undertake a remediation effort at a scale that had never before been attempted. In many ways, the mine was in uncharted territory that required new equipment, new processes, and, most of all, new thinking. To be successful would not only require the considerable talents of the people at the mine, it necessitated many independent experts to review and challenge the plans before they were implemented. This team of experts started out as an ad hoc group to look at very specific issues, but was later formalized into a group called the Mine Technical Review Team (MTRT) and tasked with the independent review of almost all critical plans. The MTRT has become a critical component in the technical process at the Bingham Canyon Mine and continued on after recovering from the Manefay.

The independent review of the work after the Manefay failure was started by a general manager in the Rio Tinto Technology and Innovation (T&I) group, after Matt Lengerich requested help in managing the decision-making process for remediation of the head scarp. As part of the process, arrangements were made for independent experts to be a part of a large team working on making a determination of when and where manned equipment could be used on the head scarp.

On May 28 and 29, 2013, the independent experts—including Mark Hawley, president of Piteau Associates; Richard Davidson, senior principal and vice president of URS Corporation (now an AECOM company); and Kim McCarter from the University of Utah (and a former Kennecott employee)—met with a group of Rio Tinto and Kennecott geotechnical and mine planning staff as well as the geotechnical consulting group to go through a risk assessment of different methods to stabilize the head scarp. Various options were presented and discussed along with the geotechnical analyses that had been completed to date.

During the meeting, several recommendations were made, such as the areas where manned equipment could be used and remote-controlled equipment was required. Other questions, such as how much of the head scarp had to be removed to make it safe, could not be answered. The independent and Rio Tinto experts needed more data and analyses before final recommendations could be made on the mining of the head scarp. Importantly, the group required the mine's Geotechnical Department to install additional monitoring with downhole measurements, such as time-domain reflectometers and downhole extensometers. The team also wanted additional sampling with both drill holes and grab samples to get a better understanding of the rock properties for the Manefay and the overall rock mass.

The other recommendation from the experts was that the geotechnical analysis needed to be completed with three-dimensional (3D) modeling in addition to the two-dimensional analysis that had been done in the past. Two types of 3D analysis programs were to be used on the head scarp, which included Clara-W, which is a 3D limit equilibrium program, and Flac3D, a numerical analysis program.

Numerical analysis programs use explicit finite formulations that model strain and displacement. These models can be used to understand large-scale and complex situations. The benefit of these types of programs is that they model more failure mechanisms than limit equilibrium programs. The downsides are the complexity, the time required to build the basic model, and the length of time needed for individual cases to run (large models can take days to calculate a result). Limit equilibrium programs basically compare forces, such as gravity, that make masses slide to the forces that resist sliding, such as rock strength and friction. Although simple compared to a numerical analysis method, limit equilibrium models have the advantage of being easier to build and the calculation time is much faster. The downside is that limit equilibrium programs do not model all types of failure modes.

During the month of June, the mine's Geotechnical Team and the geotechnical consultants were extremely busy gathering the data and building the models required by the independent and Rio Tinto geotechnical experts. This team of experts were informally called the Technical Review Team. Mark Button organized the team, which was critical to effectively utilize the group.

As the month of June progressed, questions were being formulated in regard to mining the head scarp that the Technical Review Team would be instrumental in resolving. Some of the questions included those that follow:

- How much warning should the monitoring systems provide in case of failure of the head scarp?
- What was the risk of head scarp failure from blasting?
- What blasting criteria should be used to protect the head scarp?
- What were the appropriate geotechnical modeling programs for analyzing the head scarp?
- Were the geotechnical parameters used in modeling appropriate?
- Were the operational methods appropriate to mitigate the risks?
- What factor of safety (FOS) should be achieved to allow for manned equipment on or below the head scarp?

Resolving each of these questions was critical for the Next Ore Team to be able to complete the work on the head scarp. But determining the answers was not easy, even with the independent experts in the Technical Review Team working on it. One of the most important questions was the geotechnical design criteria requirement. The original planning at Bingham Canyon Mine had been done using 2D analyses, but after the Manefay, the Technical Review Team pushed for the geotechnical analysis to be done in three dimensions. Because of that change, there were questions as to whether the 3D analysis required different design criteria because of the additional dimension.

The FOS calculation is the ratio of resisting forces in a slope (such as rock strength) to the driving forces trying to make the slope fail (such as gravity and water pressure), according to *Guidelines for Open Pit Slope Design* by Read and Stacey. If the FOS is calculated to be 1, then the slope is in equilibrium. The larger the FOS, the more stable the slope. *Guidelines for Open Pit Slope Design* recommends a range of acceptable FOS from 1.1 to 1.5, depending on the confidence level of the data used to calculate the FOS as well as the critical nature of the slope. The independent experts were critical in setting the required FOS for the remediation and planning work, which dictated the amount of work that had to be done to make the areas safe without using remote-controlled equipment.

By the end of June, the Next Ore Team had progressed far enough with the mining of the head scarp that it needed answers on what had to be mined to make the mass stable. The solution was critical because an increase of 0.15 in the FOS would change the unweighting requirement from 10 million tons to 26 million tons.

The Technical Review Team met on June 28 to debate the FOS criteria as well as a number of operational method issues. There were two strong camps, and each advocated a different criterion for the FOS. A group including the mine Geotechnical and Operations teams were promoting the lower FOS because the head scarp was being closely monitored and people would not be placed in danger because they could be evacuated, similar to the Manefay situation.

The other group advocating a higher FOS were primarily from the Rio Tinto T&I group. The basis for using a higher FOS was to stay within industry and Rio Tinto standards. The general manager for Strategic Production Planning, Technology and Innovation, joined the meeting. The general manager made strong arguments for the higher FOS and recommended breakout sessions to work on the issues. The advantage of the breakout sessions was that they were smaller groups and it would be easier to reach an agreement.

The debate went late into the afternoon, and the Technical Review Team was having a difficult time coming to a resolution. At one point, they went into an executive session with just a few invited attendees to try to resolve the issue. A final agreement could not be achieved. Finally, the team adjourned for the day, then met again during and after dinner. It was during this time that they made the most progress in reaching a conclusion.

The next morning, a recommendation was made by the Technical Review Team. Because the head scarp mass was on a weak bed that had already failed and people were working on and below the potential failure mass, the long-term criteria would be to increase the FOS by 0.15 in 3D analysis compared to the previous criteria that were used at the mine. As the mine continued to remove tons off the failure mass, the potential for a brittle failure would be reduced and monitoring would give sufficient warning to keep people safe. The team concluded that once the mine reached an FOS of 0.10 greater than the previous criteria, manned work could begin both on and below the scarp, and the No Go Zone on the head scarp could be eliminated.

The Next Ore Team now had a definitive target to reach regarding when they could start the manned work. Perhaps they would have liked to have had a lower target, but knowing the target was much better than not knowing. The Next Ore Team continued to focus on the head scarp, and by August 21 they reached the lower FOS. The ultimate FOS was not met until November 2013.

The Technical Review Team had proved to be an important aspect of safely completing the remediation work. Not only did it push the Geotechnical Team to do more and better work, it ensured that the high standards were set for the Operations group. The team was also practical enough that it could find ways to help the teams be safe while trying to meet its goals and targets.

Mine Technical Review Team

In early August of 2013, the Geology and Geotechnical Departments and the Mine Planning Department were combined into one department to create a more integrated planning group. This made sense because it was critical that these departments work closely together as each relied on the others to be successful. Too many times in the past there were silos that slowed the flow of information or created bottlenecks in the decision-making process. At that same time, Rohan McGowan-Jackson, the vice president for innovation and resource development, was pushing us to have a more formal independent assurance program. Although the Technical Review Team was doing a good job, it was only focused on Manefay geotechnical issues and did not have a formal charter.

To address Rohan's issues, both Martyn Robotham and Mark Button were recruited to help draft a charter for the Technical Review Team. Martyn made the first draft of the document, which covered the basics of who was on the team and their primary responsibilities. After the draft was completed, Mark and I began to work on modifications. The first change was to broaden the scope of the team to include the review of hydrogeology and mine planning.

In reality, the areas of geotechnical engineering, hydrogeology, mine planning, and consultants had to work together to fully cover the more technical aspects of the remediation work and future mine plans.

The name of the independent review team was changed to *Mine* Technical Review Team (MTRT) to reflect that the team would now review all technical aspects of the mine. Two new members were recruited into the team: Andrew Holmes from Piteau Associates as the hydrogeology expert and Tim Swendseid, president of the Americas for RungePincockMinarco, was brought in as a mine planning expert.

In addition to the formal members of the MTRT, the mine's Technical Services Department manager would chair the meetings but would not be a voting member of the team. However, as chairman, the Technical Services manager controlled when the MTRT would meet as well as the agenda for the meeting and therefore the topics discussed by the MTRT. The Technical Services manager also controlled the technical work being performed at the mine and therefore was responsible for the information presented to the MTRT.

To balance the accountability between the MTRT and the mine, a clause was added to the charter. This clause stated that after each meeting, the MTRT external members would send a report directly to the managing director of Kennecott that detailed the effectiveness and progress of the team and the technical work being done at the mine. This clause not only gave the MTRT more power, since they reported directly to the managing director, but it also ensured that the technical (especially geotechnical) issues remained a high priority with the company. Figure 7.11 shows the relationship between the various parties.

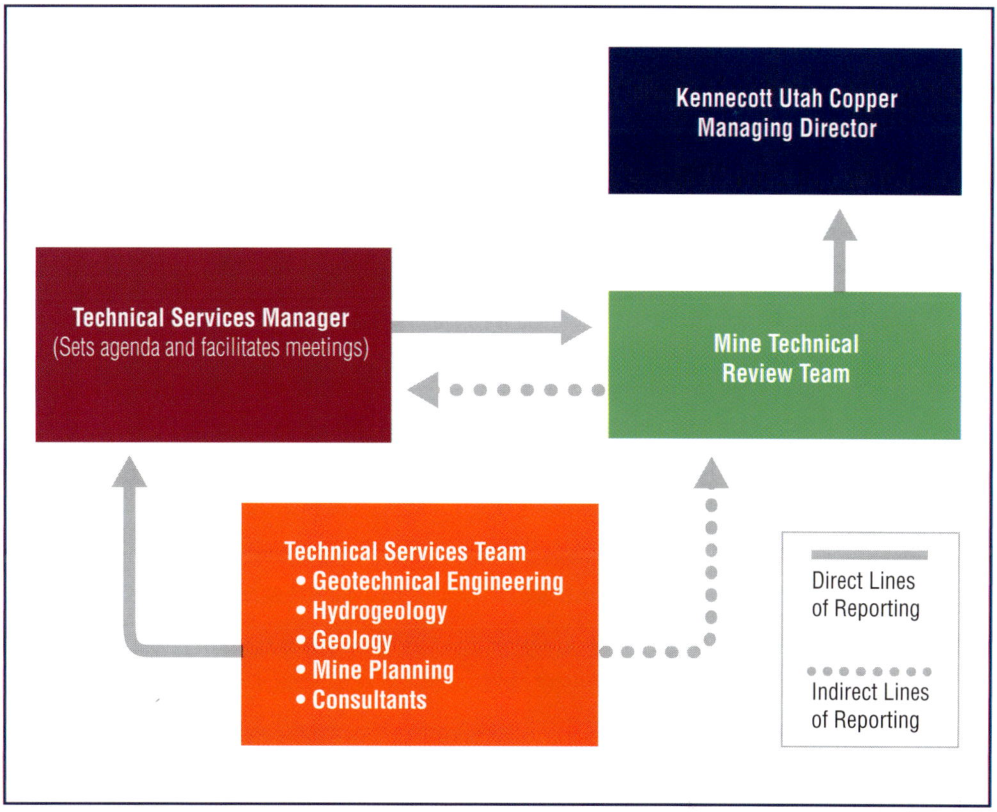

Figure 7.11 • MTRT Reporting Structure

The Technical Services Team came to rely heavily on the MTRT over the next two years. It was an important group not only because the members wielded power from their reporting relationship, but they also commanded respect because of their knowledge and skills. The MTRT challenged the Remediation and Technical Services Teams to do more and better, which forced the entire technical organization to grow professionally. Not only did presentation skills increase, the capabilities of the people improved as they began to anticipate issues and questions that the MTRT would bring forward and learned to address them ahead of time.

At the start of the MTRT process, deliberations were focused exclusively on the head scarp, which was the most critical work at the time. From there, the MTRT topics of review and recommendation continued to grow because of the complexity and uniqueness of the work being done by the Next Ore Team. Telephone conferences were held almost every other week and on-site visits at least once a quarter. Over the next 18 months, the MTRT would meet nearly 40 times to listen to presentations, review data and models, and provide feedback. Figure 7.12 shows the range of topics that the MTRT became involved in.

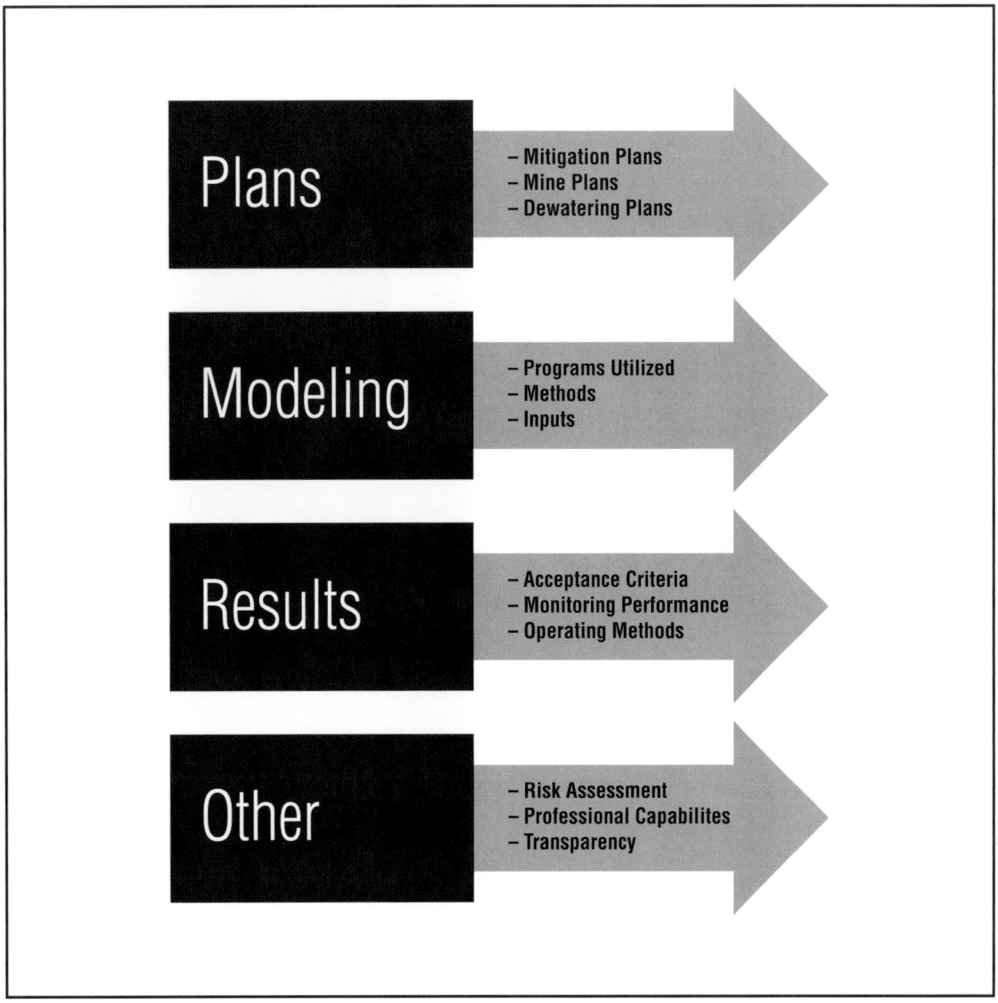

Figure 7.12 • MTRT Areas of Review

The MTRT provided a critical service to the Bingham Canyon Mine in its journey to recover from the Manefay. The team forced the staff to consider alternatives and not take shortcuts while working to a high standard. The MTRT has also shared their knowledge and experience, which has increased understanding and skills of all geotechnical employees and consultants. The MTRT members were extremely responsive—they understood the importance of time and always found a way to review and respond quickly as issues or questions were brought to them.

Before the Manefay, there was a reluctance to bring in independent reviewers because of the time, cost, or disruption that might result. After the Manefay, the MTRT proved that independent technical reviews could be effective, efficient, and critical to a successful operation.

Creating a New Culture

The Manefay was such a large crisis that the sheer magnitude would have changed the culture and the way any organization worked. How the culture would change at Kennecott was an unknown in the beginning. Would the changes be guided and planned or spontaneous and random? Would those changes be the catalyst for the company to rise to the occasion and survive—or its downfall and death? And would the changes to the culture be sustainable and drive the company going forward or be temporary and disappear as quickly as the Manefay failure itself?

Although Kelly Sanders, president and CEO of Kennecott, and Stephane Leblanc, chief operating officer, largely effected the culture change when they created the goal of returning to production in just two days after the Manefay, the reality was that the cultural changes were set in motion well before the Manefay by the people and policies that were in place beforehand. Kelly selected leaders for his management team with the qualities and capabilities that he believed could create a strong and lasting culture at Kennecott. These leaders in turn selected managers, supervisors, and hourly employees who showed the same characteristics.

Kelly has always put a lot of thought into the culture of a company and how he could influence it to be better and stronger. It was no accident that a leader like Stephane was selected to help drive change into the organization by setting aggressive goals and expecting people to deliver on them. Or Matt Lengerich, who as a young general manager had the personality to listen to a variety of viewpoints and set a clear direction for the team going forward while showing that he cared for and supported the people he worked with.

Kelly understood that the challenges of a large company like Kennecott would be met and overcome by the people—not just equipment, policies, or money. The Manefay recovery would take motivation, innovation, and teamwork, all of which required a culture that valued these attributes. Based on a vision of this culture, Kelly selected leaders with fundamental basic principles that included

- Transparency and open communication,
- Value and appreciation of people,
- Ability to work as a team, and
- High expectations for themselves and others.

Transparency and Open Communications

Although individual leaders had their own strengths and ways of doing things, there were basic principles and beliefs that were the foundation of the culture changes during the Manefay. Transparency and open communication was the highest among them. This could be seen with both the internal and stakeholder communications to government officials, such as the Mine Safety and Health Administration and elected officials, the news organizations, and the vendors that did business with the company. This transparency and communication paid off many times over because people knew the issues and were committed to remediating them. The help, support, and effort that the company received before and after the Manefay was nothing short of phenomenal. Everyone stepped up because they understood what was needed and what was at stake.

This transparency and open communication went beyond keeping the employees and stakeholders directly involved in the Manefay informed. In August of 2013, the company's senior leadership gave approval that a significant amount of information about the Manefay could be shared with the entire mining industry at the Society for Mining, Metallurgy & Exploration Inc. (SME) conference scheduled for Salt Lake City in February of 2014. At that time, the leadership did not know how successful the remediation would be, but they did know it was an important process that others in the industry could learn from. The Kennecott leadership approved two full sessions, or nearly five hours of presentations, to be given at the conference on topics that ranged from how the mine prepared for the failure to what was being done to remediate the Manefay. In previous years, Bingham Canyon Mine had limited participation in the SME conferences, so allowing two full sessions was unprecedented for the company. There was some information that could not be shared at the conference in Salt Lake City because of confidential business issues, but allowing the presentations demonstrated the leadership's commitment to communication.

Value People

The second foundation to the culture was to value people. This was demonstrated by the company's commitment to safety. Before the Manefay, the equipment was left in the pit so production could start as soon as possible, but the people were evacuated. After the Manefay, people were working a tremendous number of hours, not because they had to, but because they knew it was needed and appreciated. Whenever a member of the Senior Leadership Team would meet one of the employees, one of the first questions the employee was asked was, "How are you doing?"—not in the typical friendly and polite manner, but in a tone that demonstrated they really wanted to know: "Are you getting enough rest? Have you been exercising? Are you spending time with your family?"

For example, a few weeks after the Manefay failure, I was meeting with Rohan McGowan-Jackson, vice-president of Innovation and Resource Development, and I became a little short with one of my answers. Instead of getting mad or defensive, Rohan asked how I was doing and whether I was getting enough exercise. After listening to my answers (and excuses), Rohan made me commit to getting to the gym within the next couple of days. At the time, I thought it was a waste of time; I had too much to do. But a commitment is a commitment, so that evening I went to the gym for an hour. To my surprise that hour helped me think more clearly and put things in perspective. As a result, regular exercise again became a normal part of my routine.

Just as the Senior Leadership Team modeled their concern and value for people, these values started to spread to all levels of the company. It was not unusual for someone to ask how many hours an employee had worked in a week, and if hours worked seemed to be getting excessive, to recommend, or in some cases require, that time be taken off. Often this time off was only a weekend, but it made a difference with morale and effectiveness. Senior leadership also started to model the behavior they wanted from their employees. The first time that Matt Lengerich actually took a few days of vacation time was noticed, and people decided that perhaps it was time that they took some time off as well. It was an important message that kept employees energized and prevented burnout.

Teamwork

The ability to work as a team has a direct relationship to motivation and innovation. Kennecott used programs such as Six Sigma and Lean for years before the Manefay slide as a way to motivate the entire workforce to find ways to improve the company. These programs were the foundation for much of the work after the Manefay where employees from all levels of the company worked in teams to do the detailed risk assessments and Management of Change analyses for work the company had never undertaken before. The company became much more of a "thinking" organization that solved problems instead of rushing in and just trying to get the work done. It is this process that helped keep people safe after the Manefay. After the Manefay, the company reached *more than a million hours without a reportable injury*—a remarkable achievement considering the unique nature of much of the work.

High Expectations

Setting high expectations for the mine to quickly return to production after the Manefay was critical for the ultimate success of the remediation work and getting the mine back to normal production. However, setting high expectations for the downstream facilities at the concentrator, smelter, and refinery was completely different. The mine had more work to do than ever before and required additional resources to get it done, whereas the downstream processes would receive less than half the budgeted ore, which could cause disruptions to their normal methods and tremendous inefficiencies. The setting of high expectations for these plants meant managing the low tonnages, reducing costs, increasing recoveries, and finding alternatives to the normal ore from the mine—in essence, finding a way to meet these high expectations.

Culture Change

In the nine months from February 2013 when it was determined that the Manefay would fail, to November 13 when the Next Ore was delivered, the Bingham Canyon Mine went through a tremendous transformation—both physically and culturally. The physical changes were evident and can be seen in Figure 7.13, which shows the wall of the Manefay before April 10, just after the failure on April 11, and finally after the next ore was delivered on November 13. The impact to the mine was remarkable.

As great as the physical impact, the cultural change was even greater. Before the Manefay, Bingham Canyon Mine had been in existence for more than 100 years and, like many mature mines, the company was comfortable with its procedures and confident with its knowledge and capabilities. In some ways, this made people fairly insular and hesitant to seek outside help.

After the Manefay, the entire workforce was motivated to work together to solve problems and return the company to full production. Trust between management and labor was at an all-time high. Innovation was happening at a breakneck pace. Safety was paramount, and everyone was focused on preventing accidents and injuries. Bureaucracy was minimized, and the silos between work groups seemed to melt away.

It was a time of long hours, uncertainty of the future, and great stress to achieve seemingly impossible goals. But the reality is, the company was tremendously successful—no one was hurt or injured in the largest landslide in mining history, recovery was much faster than people believed possible, while costs were reduced and recoveries increased.

Figure 7.13 • Manefay Progression

Lessons Learned from Innovation, Technology, and Culture Change

Create the conditions for innovation. Because of the Manefay failure, there was a tremendous need to do things differently and to be innovative. The management of the company leveraged off of this need by setting difficult targets, but also gave employees the freedom (and resources) to find a way to meet those objectives. It was also an environment where people could fail (as long as it did not create a safety issue). The expectation was: if you are going to fail, do it quickly so you can try something else. Consequently, several people were given the opportunity to try new things and manage projects who would not normally have had that chance because of their position in the company or experience level. Few would say this was an easy time, but many would agree that it was a challenging and exciting part of their career that will not be forgotten.

Perform an independent review. One of the largest cultural changes before and after the Manefay was the company's willingness to use independent experts to review and challenge plans and procedures. This change helped keep people safe and greatly benefitted the recovery process by providing different perspectives and knowledge.

Create a new culture. The culture change that the Bingham Canyon Mine went through after the Manefay so the company could be more transparent, inclusive, and innovative was not accidental. The senior leadership had a vision of what was needed and put the people, resources, and motivation in place to make change happen.

Understand the needs. Although the Manefay affects all facets of Kennecott Utah Copper operations, different parts of the company had different needs, expectations, and drivers. The mine needed to return to production as quickly as possible (while still being safe), so large amounts of equipment, people, and resources flowed to the mine. At the same time, the concentrator needed to focus on recovery to get as much as possible out of the processed ore, and the smelter and refinery were focused on controlling costs. If Kennecott had managed the Manefay with a one-size-fits-all approach, the company would not have been nearly as successful.

Afterword

When I was writing this book, my greatest fears were: Would anyone read it, and if they did, would it make a difference? Those questions were partially answered in April of 2016 when the book was close to its final form and had gone through multiple edits. At that time, the latest draft was being reviewed by several people at Kennecott, including Nigel Steward, the managing director. Nigel had arrived at the mine in 2015 and therefore had not experienced the Manefay event firsthand. One of the reasons that Nigel wanted to review the manuscript was to gain a greater understanding of what had happened and what was done to prepare for and recover from the gigantic landslide.

At the time of Nigel's review, it just so happened that the mine was experiencing movement in the south highwall in a mass called Main Hill. Although this mass was more than an order of magnitude smaller than the Manefay landslide, the geotechnical engineers were worried that the Main Hill would fail and cover a portion of the 10% haul road and potentially bury the portal to the underground workings in the bottom of the mine, creating a hazard to people and equipment if the mine was not prepared. Just as with the Manefay, the geotechnical engineers provided an early warning of the potential problem, and work was initiated to prepare for a failure. Many lessons had already been applied from the Manefay, such as using an independent group of experts to review the geotechnical work, performing three-dimensional analyses on the failure mass, and communicating the change in movement with the workforce on a daily basis. Importantly, the team also created a Trigger Action and Response Plan (TARP) for the Main Hill that was similar to the Manefay Response Plan described in Chapter 2, but it was modified to account for the specific conditions of the Main Hill. The new TARP included an additional response level where part of the lower pit would be evacuated with a longer lead time to the projected failure date than that of the Manefay.

After reading the manuscript, Nigel directed the team at the mine to take additional steps to protect people and equipment based on stories in the book. In the first chapter, the manuscript described how the Manefay failure acted differently than expected, which resulted in a significant amount of equipment being destroyed. Therefore, Nigel "wore a black hat" and challenged the geotechnical engineers to further study the area that the Main Hill failure could potentially include. The geotechnical engineers performed additional analyses that resulted in a much greater understanding of the maximum extent that a failure could cover.

Also in the first chapter is a description of the TARP that described what people needed to do at different stages of highwall movement. Because of the critical nature of following the TARP, the second action that Nigel took was to dictate that under no circumstances could anyone enter an area that had been evacuated until the risk of the highwall failure had been downgraded. To emphasize the point, Nigel stated that even if the CEO of Rio Tinto directed someone to enter the area, it would not be allowed. These were strong words that clearly communicated his expectations to everyone at the mine.

Nigel also wanted to ensure that all personnel would be accounted for during an evacuation (referring to Chapter 1 when cars were still in the parking lot after the Manefay had been evacuated). He directed that a method be established to keep track of every person who entered the mine once the TARP reached Level 2 or higher. The mine's equipment operators would be accounted for with the mine's dispatch system. Anyone not operating equipment under the dispatch system would check in and out at the 10% haul road access control point.

The outcome from Nigel reading the manuscript was the extension and enhancement of an already strong system for protecting people and equipment. This response gave me a sense of relief and excitement that the story of the Manefay really had made a difference.

There is a saying in the mining industry that rules are "written in blood," meaning that people have to die or be injured for us to learn the lessons that change our behaviors. The actions for Main Hill gave me a sense of hope—hope that the power of telling stories and sharing experiences is also effective in changing behaviors, and the lessons do not have to be bloodstained.

With this in mind, I invite you to share your stories and experiences of rising to the occasion in the face of a significant risk to prevent injury and property damage, recover from a crisis, or save or improve an organization. Even if you work in a business other than mining, there are stories that are applicable to all industries. You can send me your stories by e-mail (brad.ross@risetotheoccasion.net), go to my blog (risetotheoccasion.net), or find me on LinkedIn (www.linkedin.com/in/bradross90degreellc). Please include photos or graphics if you have them. By sharing your experiences, we can all learn from the lessons not written in blood.

Appendix A

Monitoring Systems

At least 12 systems and methods are used to monitor highwall stability at the Bingham Canyon Mine today. It might seem that some of the systems are redundant and may not add value, such as two different radar systems or two different prism systems. But each system has its advantages and disadvantages. With multiple systems, the pros of one system compensate for the cons of another system and vice versa. Considering that highwall failures were identified as a major risk to the mine and because the Geotechnical Team was able to predict the Manefay using these systems shows that having multiple processes is not only prudent, it is essential. This is why the geotechnical professionals at the Bingham Canyon Mine have continued to expand on the number and type of monitoring systems to collect data. In many ways, they are pushing the envelope of the technology and capability to use the massive amounts of data collected. Keeping mine employees safe is worth the effort.

Layers of Protection

Knowing that the Manefay bed was going to fail made the difference in having zero fatalities instead of multiple casualties. Given that the risk of highwall failures is so high, tremendous resources are invested in multiple systems to identify and predict failures before they happen at the Bingham Canyon Mine. Over time, the number and sophistication of the highwall monitoring systems have increased, which reflects the increasing risks as the Bingham Canyon Mine continues to grow wider and deeper. Before the Manefay failure, there were 10 types of monitoring systems, each adding a layer of safety. Since the Manefay, two additional monitoring systems were added (GPS prism system and downhole inclinometers). Following are the various types of systems that are employed at the Bingham Canyon Mine since the Manefay to protect people from highwall failures:

1. Geotechnical Hazard Training
2. Routine, Documented Inspections
3. Prism Network
4. Extensometers
5. Time Domain Reflectometry
6. IBIS Slope Stability Radar
7. GroundProbe Radar
8. Microseismic Monitoring
9. Geographic Information System Data Display
10. Piezometers
11. Global Positioning System Prism System
12. Downhole Inclinometers

The monitoring systems range from training personnel to high-technology radar systems that continuously monitor the highwalls. The following sections describe these systems and how they are used at the Bingham Canyon Mine.

Geotechnical Hazard Training

Every person that works at the Bingham Canyon Mine is trained to watch for geotechnical hazards and report them to the appropriate person if such conditions are observed. People are trained to observe and report

- Falling rocks,
- Cracking in highwalls and benches,
- Dust clouds from rolling rock (Figure A.1), and
- Newly fallen rock on benches (Figure A.2).

Figure A.1 • Dust Cloud

Figure A.2 • Fallen Rock on Benches

Mine Safety and Health Administration (MSHA) regulations require annual training for these hazards. At the Bingham Canyon Mine, this training is provided by the Geotechnical Department and is considered one of the most important topics by both employees and management.

Three reporting methods are used at the mine if an employee observes what he or she believes is a geotechnical hazard:

1. If the hazard appears to be a risk that can imminently injure people or damage equipment, then anyone can call a "Mayday" on the company radio, and the area or even the entire pit is evacuated.

2. If the hazard appears to be localized but has the possibility that a part of the pit should be shut down, then the person contacts Production Control so equipment can be routed away from the area, and the Geotechnical Department is notified to investigate the issue. These areas become No Go Zones that cannot be entered until the geotechnical engineers have inspected the area and determined that it is safe to return to work in the vicinity.

3. If an employee observes an issue but is not sure of its criticality, he or she then calls a supervisor. The supervisor determines the urgency and whether or not the area is to be evacuated, and the Geotechnical Department is informed.

The training of operations personnel and the resulting observation and reporting methods are taken seriously at Bingham Canyon Mine. There have been several situations where operators had observed bench-scale hazards before any of the monitoring systems detected the danger; their reporting was the earliest warning of potential hazards.

Geotechnical hazard training was especially important as the Manefay failure approached. Every person going into the pit paid even more attention to the highwalls. Having all of those eyes monitoring the slopes supplemented the geotechnical continuous monitoring. The operators reported cracks in the 10% haul road the day before the Manefay failure, which increased the level of concern about the upcoming failure.

Routine, Documented Inspections

On a routine basis, both supervisors and the geotechnical engineers inspect the pit for hazards. Pit supervisors inspect work areas before every shift, as required by MSHA, and these inspections are documented in a logbook. When the supervisors observe a potential geotechnical hazard, they notify the Geotechnical Department. The supervisors will also move people and equipment out of areas if warranted.

Geotechnical engineers inspect the pit on a daily basis looking for issues in both immediate working and inactive areas of the mine to check for changes in conditions. The engineers combine these inspections with other monitoring data to develop a comprehensive understanding of the geotechnical situation in the entire pit. On a monthly basis, a slide hazard map is generated and distributed to operations and mine planning personnel. Operations personnel post these maps in crew line-out rooms so operators know where the potential hazard areas are and can be especially aware of geotechnical changes. The mine planners use the map to design potential remediation work and schedule equipment for the following month.

If significant geotechnical hazards are observed, an action plan is developed. After ensuring that the immediate hazards are taken care of, the geotechnical engineer will put together a "Geotechnical Bulletin" that describes the issue and action items to keep people safe and remediate the issue. This bulletin is sent to all mine personnel and posted in crew line-out rooms. A copy of the Geotechnical Bulletin that was published on February 14, 2013, is shown in Figure A.3.

Rio Tinto

GEOTECHNICAL BULLETIN

Distribution: Mine Area Leaders, Superintend and above.

Subject: Deformation along the Manefay Fault, Eastside pit wall.

Date: February 14, 2013

Originators: Mine Geotechnical Department

Advisory: Potential Geotechnical Hazard - East Wall, Manefay Fault.

Description:

An area of highwall on the East wall of the pit, associated with the Manefay Fault, has shown low levels movement averaging 0.1"/day. Tension cracks associated with this movement have developed adjacent to MMC and the Visitor Center. At this point there is no risk to mine personnel.

IBIS and GroundProbe radars are currently scanning the area and outputs from these instruments are being monitored for changes in movement rates. Prism surveys and extensometers are also supporting monitoring efforts. Geotechnical staff is performing daily inspections in the affected areas to track changing conditions. In addition, contingency plans are being developed in the event that any changes in slope condition(s) may require further actions to assure the safety of Mine personnel.

Action Items:

- Ground control monitoring – radar units, prisms, extensometers, and other monitoring systems are in-place and provide continuous on slope conditions. Monitoring systems are alarmed and responded to 24 hours per day. Geotechnical staff will continue to monitor and provide information to the Mine in the event of any change.
- Mine personnel should be aware of changes in highwall conditions and report these to their supervisor and/or geotechnical staff.
- Remote location testing for Mine Monitoring and Control (MMC) facilities and personnel will be initiated to demonstrate a proven capability for continued operation should changes in slope movement dictate the necessity for relocation.
- Contingency planning for any other impacts resulting from a change in slope movement rates will be developed as needed.
- Further communications on the status of changes in the conditions associated with the Manefay fault, will be distributed to Mine personnel.

Please direct any questions or observations immediately to your leader

Figure A.3 • Geotechnical Bulletin for February 14, 2013

RioTinto

Figure 1. Movement zone above the Manefay Fault.

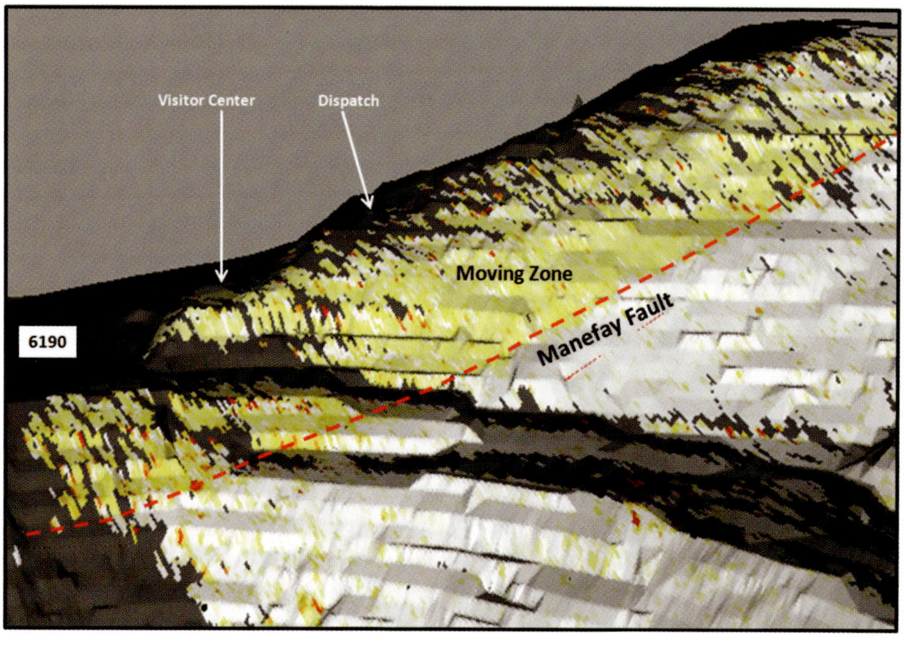

Figure 2. Movement zone (above the Manefay Fault) as seen by IBIS radar.

Figure A.3 • Geotechnical Bulletin for February 14, 2013 (continued)

Prism Network

The prism network uses a series of prism targets that are attached to the highwall. Before the Manefay, there were more than 300 of these targets placed throughout the Bingham Canyon Mine. Five robotic theodolites at fixed locations are used to accurately measure the distance to these targets to within 1/1,000th of an inch, depending on the distance to the target. By tracking the changes in distance, the geotechnical engineers can trace movement of the highwall. Figure A.4 shows the robotic theodolite used to measure distance.

Figure A.4 • Prism Network

The theodolites automatically locate each target and record the distance. These data are then automatically uploaded to a central database. Geotechnical engineers can track current movement as well as changes over different time periods. The frequency that the measurements are taken is dependent on the number of targets that the theodolites must measure. With a large number of points, the frequency of measurement for each point may be more than an hour.

Compared to extensometers that measure the width of cracks in the surface, prisms are much easier and faster to install. The prism network does not rely on cracks to be present to measure movement but instead measures the distance of a number of points over a large area.

The prism network is one of the key monitoring systems used at the mine before and after the Manefay. The data collected from the prisms is used to determine longer-term trends and movements to predict when a failure may actually occur.

Extensometers

Before the Manefay failed, several large cracks were observed behind the highwall in the area of the Visitors Center that overlooked the Bingham pit. These cracks were mapped with their location, orientation, and size. The maps and radar data were critical to confirm the movement of the Manefay mass. Some of the mapped cracks were

Figure A.5 Surface Crack

Figure A.6 • Extensometer

monitored with extensometers that continuously monitor the width of the cracks and detect whether they were moving or growing, an indication that a slope is shifting. Figure A.5 shows an example of a crack on the edge of a highwall of the Bingham Canyon Mine.

An extensometer is a simple device where a wire line is anchored on one side of the crack and connected to a supply spool that is anchored on the other side of the crack. As the crack widens, the length of wire is continuously measured to determine the amount of movement. Extensometers are attached to telemetry systems that automatically take readings and transmit the data to a central database.

The advantage of extensometers is that they are relatively simple and monitor a point of a known problem. The downside is that they measure a single point and highwall failures can occur over a large area, so the single point may not be representative or detect changes away from the point. Changes in temperature can also create false readings of movement because the wire can expand or contract with changes in temperature.

The movement rates are automatically alarmed, so movement information with sudden acceleration would immediately be relayed to dispatch and operations. Other monitoring systems, such as the prisms or radars, gather the movement periodically, and then the movements have to be calculated based on previous measurements. This results in a delay in reporting the movement that can range from a few minutes to an hour. The almost instantaneous nature of the extensometers provides the earliest warning possible. Figure A.6 shows an extensometer at one of the cracks that occurred before the Manefay failed.

Time Domain Reflectometry

Observations, extensometers, and prism networks are systems to monitor surface movements. Although understanding the surface movement is critical, highwall failures have a three-dimensional component that is not observed on the surface. Time domain reflectometry, or reflectometer (TDR), is a method to determine if there is movement below the surface and the depth of that movement.

With TDR, a transmitter sends an electrical pulse down a coaxial cable. If there is a change in the properties of the cable (such as distortion from the cable being stretched), a portion of the electrical pulse is reflected back toward the TDR transmitter. Similar to radar, TDR measures the time that the pulse is reflected back and is then able to calculate the distance of the distortion from the transmitter (Benson and Bosscher 1999). For geotechnical monitoring, the coaxial cable is connected to a pipe or casing so it can be lowered down a drill hole. Once in place, movement in the underground rock will cause the coaxial cable to deform or break, which results in a distortion of the cable. The TDR can then be used to determine the distance to the movement from the surface. As long as the cable is not severed, the TDR can be used to detect the location of multiple distortions in the cable.

Ultimately the TDR measures impedance (effective resistance) in the coaxial cable and does not measure the amount of movement in the rock itself. By taking measurements at various times, the geotechnical engineers can determine if the rock mass is moving by changes in the impedance, but not the amount or direction of movement. Figure A.7 shows seven readings taken from January 12 to May 9, 2014. The colored lines are the TDR readings, and the vertical axis is the depth of the readings from the surface. As can be seen in the graph, there is some additional impedance measured at the top of Bed 1 starting on February 25. This indicates the start of some movement in the rock mass. The impedance continues to increase with each reading. Unfortunately, the amount of movement cannot be related to the change in impedance; it only indicates that the movement is increasing.

A second change in impedance of 36 is observed in the measurements on the top of Bed 2 on May 2. The impedance reduces slightly to 35 on May 9, which indicates that the movement stopped.

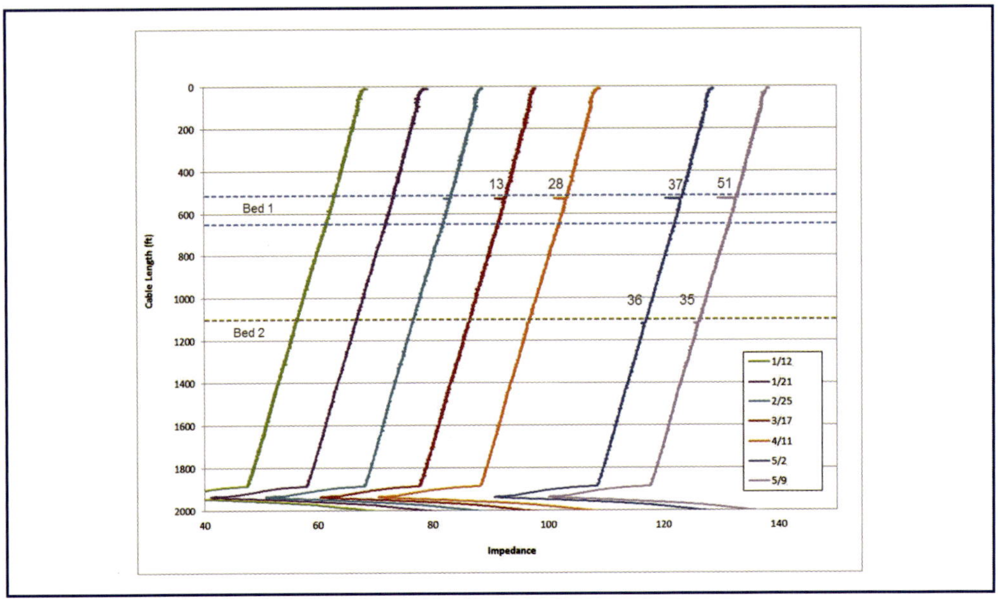

Figure A.7 • TDR Example

Disadvantages for TDR monitors are that they can only indicate movement along a single line. They cannot indicate the amount of movement or movement across a large area. The final disadvantage of TDRs, at least at the Bingham Canyon Mine before the Manefay, was that readings were taken manually, so movement would only be detected when an engineer or technician physically went to the drill site and took the reading. Readings occurred weekly, or perhaps daily if it was an area of concern. TDR cannot be used to warn people of an impending failure, but it can be used to understand the failure mechanism.

There were no operational TDRs in the Manefay mass before the failure. Drill holes were in the process of being drilled for the TDRs when the Manefay failed.

Since the Manefay, the mine has installed a much larger network of TDR monitors, and the data from these holes is sent to geotechnical engineers through telemetry instead of manual readings. Figure A.8 shows how solar panels are used to power the telemetry systems that send the data gathered from TDR and piezometer drill holes to the central database.

Figure A.8 • TDR and Piezometer Drill Holes with Telemetry

IBIS Slope Stability Radar

The IBIS radar system used at the Bingham Canyon Mine is manufactured by IDS (now part of Hexagon). These systems are semi-permanent at the mine in that they are in fixed locations for long periods of time, but they can be moved if required. The advantage of the IBIS system is that it can monitor a large portion of the open pit from a long distance away. Even though the Bingham Canyon Mine is 2½ miles in diameter, the IBIS radar devices can be set up on one side of the pit to monitor the highwalls on the far side. The devices are line of sight, so an obstruction can shade the monitoring in an area. Figure A.9 shows an IBIS radar system located on the west side of Bingham Canyon Mine and the corresponding area of the highwall that the system monitors.

Before the Manefay, there were three IBIS radars continuously monitoring the highwalls on approximately three-fourths of the mine, which included the Manefay area. Since the Manefay, an additional IBIS radar has been purchased so that all 360 degrees of the pit are monitored with the system. Radar systems do an excellent job of detecting movement in the highwalls. The data can be used to show the amount of movement over any time period for any area of the mine that has coverage. The IBIS radar system was significant in identifying the Manefay failure. With the radar, a distinct difference in movement could be seen in the rock mass above the Manefay bed compared with a much lower movement below the bed. It was this difference in movement, and the fact that the difference was accelerating, that convinced the geotechnical engineers that a large-scale failure would occur.

The radar systems are great for seeing the "big picture," but they do not totally replace other systems at the Bingham Canyon Mine. Radar measurements are influenced by changes in temperature or moisture. The greater the distance, the more the variation in measurement. When the data are plotted, there is a clear variation in movement, even based on the time of day. This can be important when very small changes are critical for calculating movement rates.

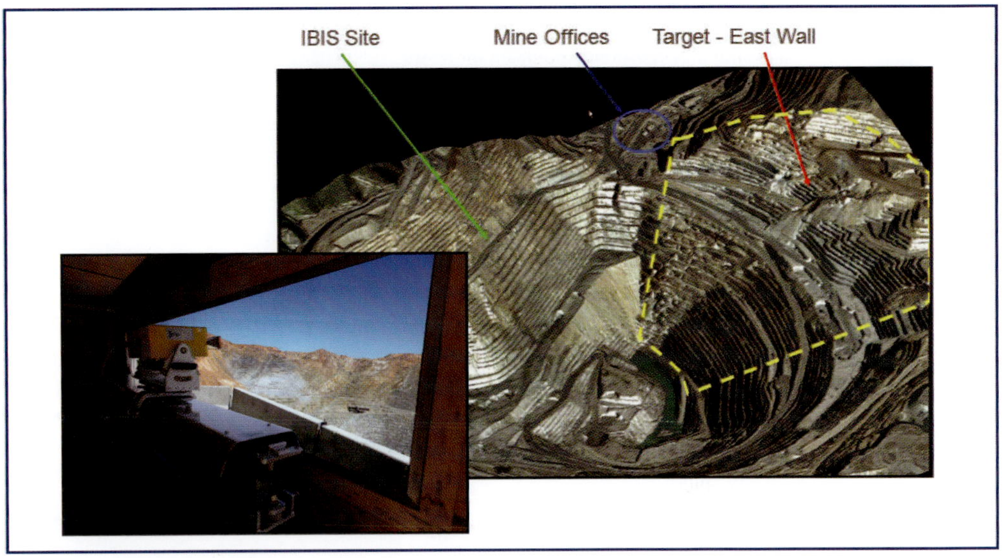

Figure A.9 • IBIS Radar System

GroundProbe Radar

A second radar system used at the Bingham Canyon Mine is made by GroundProbe out of Australia. GroundProbe radars are trailer mounted and portable. At the mine these radars are used to monitor areas that may have shown some instability or movement. Once an area of concern is identified, a GroundProbe radar is moved as close as possible without interfering with operations, which may be a few hundred to a few thousand feet away. The radar is focused on the specific area and is used to alert operations if the area becomes unstable or starts to fail. The GroundProbe radars are excellent for monitoring the normal bench-scale instabilities, and that data can then be used to predict when those failures might occur. Figure A.10 shows a GroundProbe radar system monitoring the Bingham Canyon Mine after the Manefay slide.

Figure A.10 • GroundProbe Radar System

As with the IBIS radar system, the data from the GroundProbe radars are stored in a database and can be used to look at movement trends over time. The Bingham Canyon Mine used three GroundProbe radar systems before the Manefay and has placed two additional systems in service since the slide. After the Manefay movement had been detected, at least one of the GroundProbe radars was focused on the Manefay mass at all times.

Microseismic Monitoring

A microseismic monitoring system uses geophones or accelerometers that detect seismic energy or the sounds of moving rocks underground. These geophones or accelerometers are placed in drill holes and require multiple holes and sensors to be able to locate the source of the seismic energy (Hardy 2003). Once the seismic energy is detected and if the location can be isolated, the size of the failure mass can be determined and potentially the timing of the failure predicted.

A microseismic system was in place at the Bingham Canyon Mine before the slide, but it was on the opposite side of the mine from the Manefay failure. Although the system picked up events around the Manefay mass, those occurrences have not been directly linked to the failure event. Microseismic monitoring takes a significant amount of post-measurement calculation to determine if there has been movement. Because of the delays involved with post-measurement calculation, at this time it is not possible to use this system as a method to warn operations personnel of an impending event. Although there are problems with the microseismic systems, the Bingham Canyon Mine has installed a second system near the Manefay head scarp in the hopes of understanding and solving these problems so that microseismic monitoring can be a more effective tool.

Geographic Information System Data Display

A geographic information system (GIS) is not a monitoring system, but it serves as one of the layers of protection at the mine because it gives the geotechnical engineers the ability to effectively access and analyze the data from the monitoring systems. It is one thing to gather massive amounts of geotechnical monitoring data with multiple systems, but something else to actually use the data to protect people. With the exception of the manual inspections and TDR data (at least before the Manefay), all of the data captured by the monitoring systems is automatically sent to a centralized database with telemetry as part of a GIS at the Bingham Canyon Mine.

With the GIS data display, the geotechnical engineers are able to analyze the data from the radar systems by both location and time frame. For instance, they can look at the monitoring data for a small 100-foot-by-100-foot area or zoom out to the entire pit slope that may be 2,500 feet by 2,500 feet or anything in between. They can also look at trends over time that could range from a few hours to a number of years for any of the monitoring systems that are stored in the GIS. The GIS is a powerful tool that significantly aided the geotechnical engineers in both identifying the Manefay mass and predicting when the mass would fail.

The other value of the GIS data is that it is available to the geotechnical engineers at any place with an Internet connection. The mine does not provide 24/7 coverage by geotechnical engineers, but there is always one on call. If there is an alarm of highwall movement with one of the monitoring systems, the geotechnical engineer will be notified by text from the system as well as a phone call from the operations dispatchers. The engineer can then bring up the monitoring data from multiple systems to determine what has happened and advise operations what should be done. Having all of the data at the fingertips of the geotechnical engineers is an amazingly powerful tool that turns volumes of data into useful information.

Piezometers

Piezometers are devices placed down a drill hole to measure the elevation of the water and amount of water pressure. The existence of water and amount of water pressure in the highwall of a mine are critically important to highwall stability. High water pressure will make a highwall less stable, and the presence of water on a geologic bed or fault with a lot of clay will reduce the friction on the surface and increase the likelihood that the two surfaces will slide past each other and fail. It was believed that there was little water on the Manefay bed, but there was modest data from piezometers in the Manefay mass to confirm that belief because of a lack of drill holes in the mass.

Since the Manefay, much more emphasis has been placed on gathering piezometric data as well as dewatering the highwalls of the Bingham Canyon Mine. There are only a few options for improving highwall stability, but dewatering is probably the least costly. Another alternative is to reduce the highwall angle by mining overburden but at a high cost. The other option is to leave a buttress at the bottom of the highwall, thus decreasing ore to be mined and significantly reducing the revenue generated.

Monitoring of the water pressure with piezometers is an important monitoring system to maintain mine stability. Typically, the water levels are only monitored on a weekly or monthly basis, but since the Manefay, some of the readings are automatically sent to the mine hydrologist through the telemetry system.

GPS Prism System

After the Manefay, remediation started on the head scarp left by the gigantic slide. Existing theodolites could measure the amount of movement on the face of the highwall but could not measure the topography where the theodolite did not have a line of sight. In addition, the existing system detected movement, but not the actual

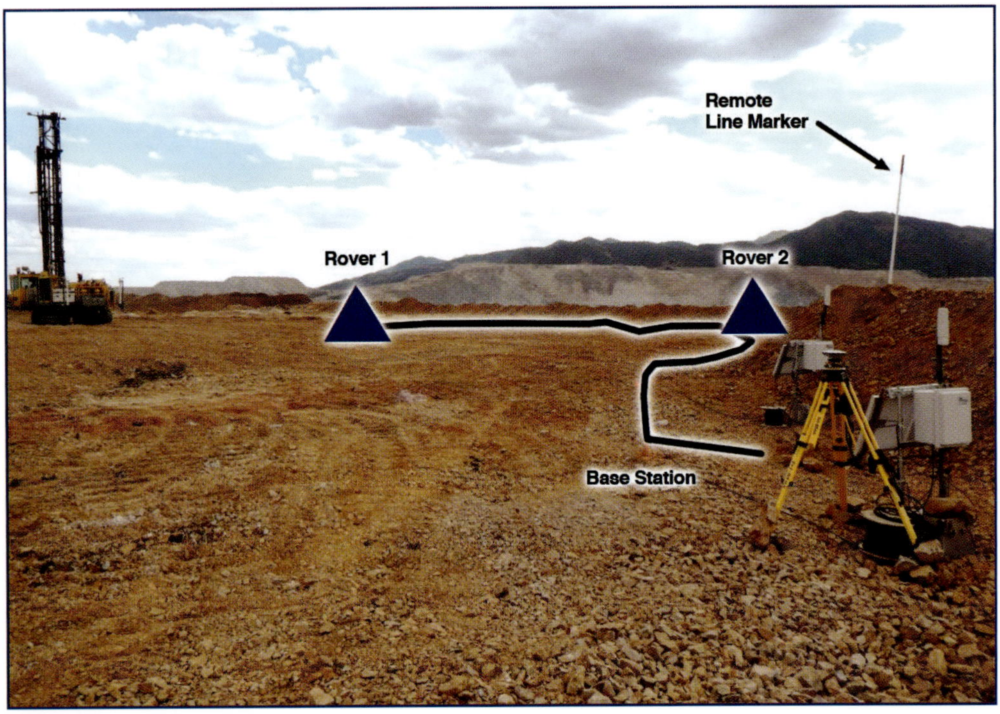

Figure A.11 • GPS Prism System

direction of the movement. This vector was important for understanding the mechanism of that movement, which could help the geotechnical engineers plan how to manage a failure.

After the Manefay, a Global Positioning System (GPS) prism system was purchased. Instead of theodolites, this system uses a GPS sensor to track movement inside of a prism. Because the mine required a great deal of accuracy, the GPS prism system was set up using a base station attached to three GPS roving prisms. The base station was used to adjust the accuracy of each of the roving prisms because it stayed in the same position and could track inaccuracies from the GPS readings. These data were also sent to the geotechnical GIS data display through telemetry as well.

With the GPS prism system, the geotechnical engineers were better able to understand the movement vectors of the highwalls. This was particularly important later in 2013 as the team worked to understand the movement on the Fortuna highwall. Figure A.11 shows a GPS prism system with two rover prisms that were set up on the Manefay head scarp.

Downhole Inclinometer

After the Manefay, the Mine Technical Review Team recommended that the geotechnical engineers evaluate the use of a downhole inclinometer. Downhole inclinometers have an advantage over TDRs in that they provide data on the amount and direction of movement. This is critical information to understand what is happening with a moving mass and the failure mechanism.

The problem with a downhole inclinometer compared to TDR is that it only measures the movement at a point in the drill hole or, in some cases, a few points. TDR measures the entire length of the drill hole. Both systems provide important information to the geotechnical engineers.

Manefay Timeline

February 2013 – June 2014

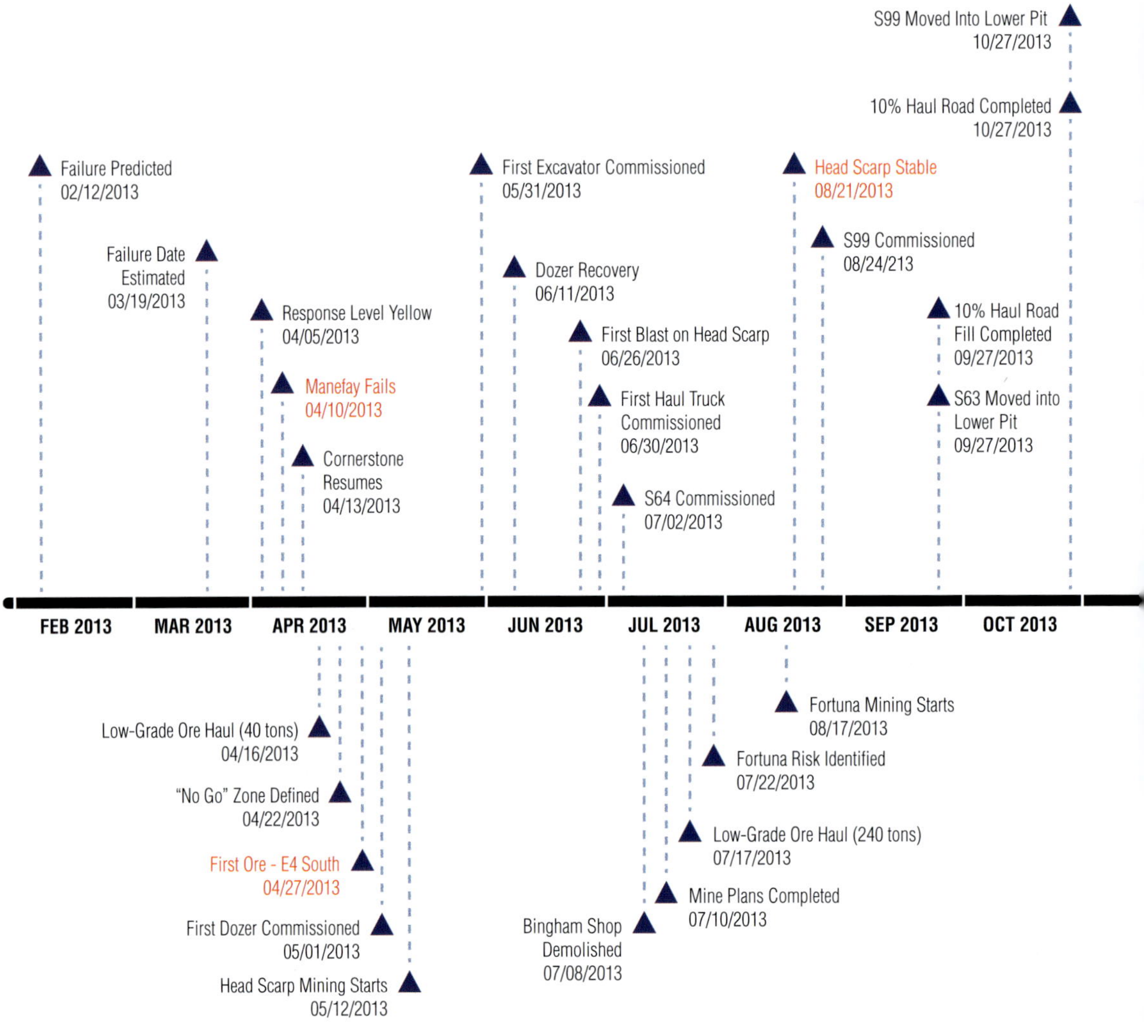

S63 = No. 63 Shovel, Hitachi 5500
S64 = No. 64 Shovel, Hitachi 5600
S99 = No. 99 Shovel, P&H 4100 XPC AC

Appendix B

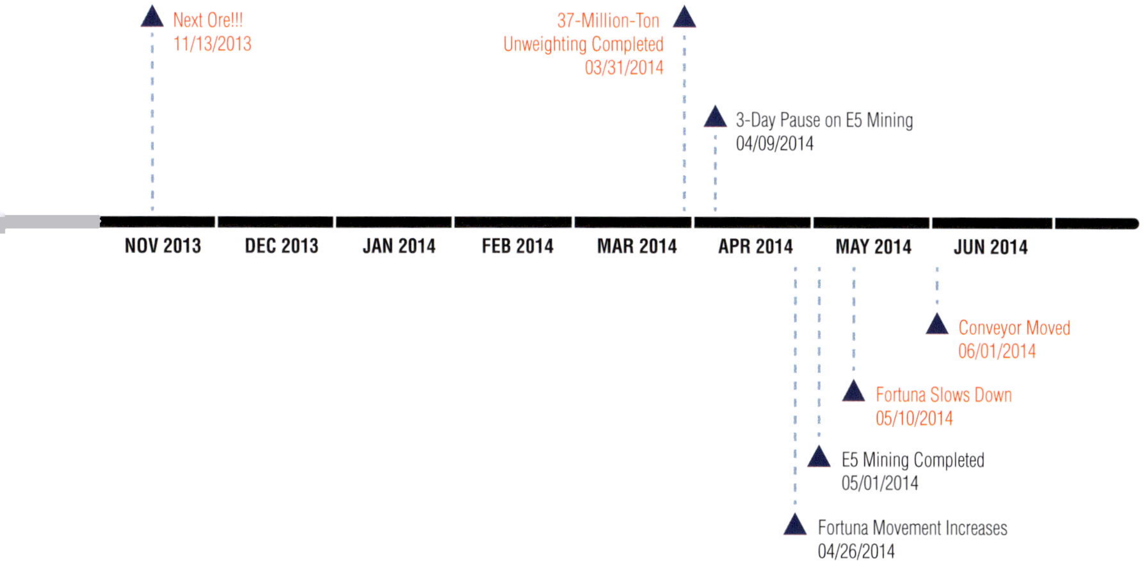

Next Ore!!!
11/13/2013
37-Million-Ton
Unweighting Completed
03/31/2014
3-Day Pause on E5 Mining
04/09/2014
NOV 2013
DEC 2013
JAN 2014
FEB 2014
MAR 2014
APR 2014
MAY 2014
JUN 2014
Conveyor Moved
06/01/2014
Fortuna Slows Down
05/10/2014
E5 Mining Completed
05/01/2014
Fortuna Movement Increases
04/26/2014

References

Benson, C.H., and Bosscher, P. 1999. Time-domain reflectometry (TDR) in geotechnics: A review. In *Nondestructive and Automated Testing for Soil and Rock Properties.* Edited by W.A. Marr and C.E. Fairhust. West Conshohocken, PA: ASTM International. pp. 113–136.

De Bono, E. 1999. *Six Thinking Hats.* New York: Back Bay Books.

Hardy, H.R. 2003. *Acoustic Emission/Microseismic Activity, Volume 1: Principles, Techniques and Geotechnical Applications.* Lisse, Netherlands: Balkema.

Pankow, K.L., Moore, J.R., Hale, J.M., Koper, K.D., Kubacki, T., Whidden, K.M., and McCarter, M.K. 2014. Massive landslide at Utah copper mine generates wealth of geophysical data. *GSA Today* 24(1):4–9.

Porter, J., Schroeder, K., and Austin, G. 2012. Geology of the Bingham Canyon porphyry Cu-Mo-Au deposit, Utah. In *Special Publication 16.* Edited by J. Hedenquist, M. Harris, and F. Camus. Littleton, CO: Society of Economic Geologists. pp. 127–146.

Read, J., and Stacey, P. (eds.). 2009. *Guidelines for Open Pit Slope Design.* Boca Raton, FL: CRC Press.

Taleb, N.N. 2010. *The Black Swan: The Impact of the Highly Improbable.* New York: Random House.

Index

Note: *f.* indicates figure; *t.* indicates table

D

E

F

G

H

I

N

O

P

R

S

T

U

V

W

Y

Z

About the Author

Brad Ross, a licensed professional mining engineer, has integrated a variety of professional, personal, and academic experiences and skills to find unique solutions to difficult problems and manage challenging projects.

After graduating from the South Dakota School of Mines in 1980, Brad became a mining engineer with Shell Mining Company, where he developed an in-depth understanding of mining through his technical work in designing new mines and production work in operating mines in Ohio and Wyoming. During this time, Brad owned and managed a small fossil business, catering to customers from around the world. This combination of technical, operational, and business experiences served him well as he moved into a variety of management positions.

Brad has managed the technical departments in some of the largest mines in the world, including Antelope and Cordero Rojo coal mines in Wyoming, the Rössing Uranium Mine in Namibia, and the Bingham Canyon Mine in Utah. Other assignments included financial controller for the Buckskin Mine in Wyoming and building a team to develop an integrated budgeting model for the five mines owned by Kennecott Energy. Brad also worked with communities and government agencies to purchase water for Resolution Copper Company in water-thirsty Arizona. Each of these jobs had their own critical issues and expanded Brad's abilities in leading people and managing projects. He became known as a person who could tackle difficult and unusual problems, which resulted in more opportunities to solve problems.

While living in Namibia, Brad and his family traveled throughout the country to meet with and purchase minerals from artisanal miners. Brad studied their living and working conditions, which led to his dissertation on the sustainability of artisanal miners in Namibia. These experiences led him to pursue his masters of engineering degree (2009) and doctoral degree (2011) in mining and geological engineering from the University of Arizona while working full time for Resolution Copper Company.

An integration of these positions and lessons learned seemed to culminate with the largest challenge of all—managing the technical services and planning efforts to prepare for and recover from the unprecedented Manefay event at the Bingham Canyon Mine in 2013. Brad then led the team charged with developing the geotechnical studies and mine planning to demonstrate the long-term viability of the mine.

In 2015, Brad left corporate life and moved to Tucson, Arizona, where he and his wife Linda enjoy the warm weather and have started the next phase of their lives. Not content to sit back and relax, Brad has become an advisor, author, and professor. He established a small consulting firm, 90 Degree Consulting LLC, that provides management advice, technical reviews, expert witness services, and mentoring of high-potential employees. Brad also gives talks about the Manefay and is writing a memoir about growing up in a small town in South Dakota, where his family lived in the county jail building—just a normal part of life when your dad is the sheriff in a remote county on an Indian reservation. In addition, Brad teaches in the Department of Mining and Geological Engineering at the University of Arizona to help cultivate the capabilities of the next generation of mining professionals.